Signal recovery from noise in electronic instrumentation

Signal recovery from noise in electronic instrumentation

T H Wilmshurst

Reader in Electronics, University of Southampton

Adam Hilger Ltd
Bristol and Boston

British Library Cataloguing in Publication Data

Wilmshurst, T. H.
  Signal recovery from noise in electronic instrumentation.
  1. Electronic instruments   2. Electronic noise
  I. Title
  621.38′0436   TK7878.4

  ISBN 0-85274-783-7

Consultant Editor: **Mr A E Bailey**

Published by Adam Hilger Ltd
Techno House, Redcliffe Way, Bristol BS1 6NX
PO Box 230, Accord, MA 02018, USA

Typeset by Mid-County Press, London SW15
Printed in Great Britain by Page Brothers (Norwich) Ltd

Contents

Preface

The subject of this book is the recovery of signals from noise in electronic instrumentation. The term 'electronic instrumentation' is sometimes reserved for instruments such as the oscilloscope and testmeter which are used to measure specifically electrical variables. It should therefore be made clear that the scope of the book goes much beyond this and covers instrumentation for the measurement of any variable, whether electrical or not. The term 'electronic' simply implies that electronic techniques are used in the process of measurement.

The term 'noise' also requires clarification. Until recently, it was used mainly to refer to random fluctuations such as white and $1/f$ noise. Now, however, the term is used to refer to almost any kind of unwanted signal in an electronic system. It is in the broader sense that the term is used in the title, since the recovery of the required signal from many kinds of unwanted signal is covered. These comprise offset, drift random noise, comprising white and $1/f$ noise, and interference. However, to avoid confusion, the term 'noise' will be reserved for random noise. The other types of unwanted signal will be referred to by their specific names.

There are, broadly, two ways of reducing the errors due to unwanted signals in an electronic system. The first is to prevent the unwanted signal from being introduced. This is a matter of low-noise amplifier design, screening, decoupling, etc, and is reasonably well covered by existing texts. The second approach, which is taken up when the first has been exploited as far as possible, is to devise 'signal recovery' techniques which distinguish the required signal as well as possible from whatever unwanted signals may remain. This second approach is less well covered at present and therefore represents the subject matter of the book.

The text is intended for anyone who is to be involved in the development of electronic instrumentation, at graduate level or beyond. In order to indicate better the intended readership it will be helpful to consider briefly the way in which electronic instruments are, or should be, developed. A number of different disciplines is nearly always involved. Consider, for example, such a system as the laser anemometer. This is a device which uses both electronic and optical techniques for the measurement of fluid velocity. Thus the development of this instrument would require experts in optics, electronics and fluid dynamics. When such a team is formed, it is important that each member should attempt to master all aspects of the system, not just those of his own discipline. This is as true for the electronic aspects of the system as for any other. Thus, the book has to be acceptable to two rather different classes of reader. The first is the undergraduate electronics engineer who intends to specialise in instrumentation. He will meet the subject probably in the third year of his undergraduate studies and will need to obtain a good grasp of the entire contents. The other class of reader will be the non-electronics specialist who will probably come to the subject after graduation and at the time of joining the team. He will usually need to study in depth a limited selection of topics from the book. In order to help this class of reader, attempts are made to cover the material in descriptive as well as analytical manner. Also, rather more extensive subsection titling than usual has been adopted, in order to facilitate 'random access'.

The book stems from two courses that are given in the Department of Electronics of the University of Southampton. The first has run for over a decade and is intended for graduates in disciplines other than electronics who are engaged in the development of electronic instrumentation, or sometimes just in its use. These may be students working for higher degrees, postdoctoral research fellows, technicians and, sometimes, academic staff.

The second course is more recent, and is intended primarily for third-year undergraduates in electronic engineering. It covers that material outside the 'core' electronics subjects that is needed for a graduate who is to specialise in electronic instrument development. The course is also followed by students taking 'with-electronics' courses such as 'physics-with-electronics', 'chemistry-with-electronics', etc.

T H Wilmshurst

1 Low-pass filtering and visual averaging

1.1 OVERVIEW

Before discussing low-pass filtering and visual averaging, the two signal recovery methods to be discussed as the main topic of the present chapter, we devote the first section to an overview of the entire book. Here the classes of unwanted signal, and also all of the signal recovery methods to be described, will be briefly introduced.

Resistor bridge strain gauge

The first step will be to describe the operation of a simple example of an electronic instrument: the resistor bridge strain gauge of figure 1.1. Here the variable to be measured is the strain (extension) of the mechanical component shown, such strain usually resulting from a mechanical force (stress) applied to the component.

The principal component of the strain gauge is the resistor R_g. This is a mesh of fine wires which is attached to the mechanical component by an adhesive. As the mechanical component is strained, R_g increases; this causes the bridge to become unbalanced, producing a voltage v_t which is approximately proportional to strain. v_t is then amplified, low-pass filtered and displayed on a chart recorder.

The variety of instruments of this kind is vast, the single common feature being that some variable, which is usually non-electrical, is converted to an electrical signal which is processed and displayed. The device which converts the variable of interest to the electrical signal is called the input 'transducer' and for the present system the transducer is the resistor bridge. This converts the variable of interest, the strain, into the electrical signal v_t. Although the range of input transducer types in use may constitute an important topic for

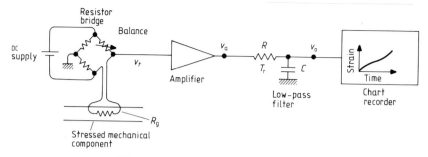

Figure 1.1 Resistor bridge strain gauge.

the instrument engineer, this is not the subject of the present book, and is adequately covered elsewhere. Instead, our objective is to establish the principles whereby the electrical signal developed by such a transducer is recovered from the various kinds of unwanted signal.

This objective will be best served if, wherever possible in the course of introducing the signal recovery methods, the same simple example of the strain gauge is retained. This is in contrast to the alternative approach of constantly altering the example in order at the same time to try to give some idea of the differing transducer types. With the principles of signal recovery thus firmly established, using one type of transducer, it will be a simple matter to apply the principles to any other type of transducer that might be encountered.

Low-pass filtering of white noise

The purpose of the low-pass filter in the strain gauge of figure 1.1 is to reduce the amplitude of the unwanted signal components, of which the first to be considered is white noise. Figure 1.2 illustrates the way in which the low-pass filter reduces such noise. Here a fixed stress is applied to the mechanical component at time $t = 0$ and the requirement is to measure the difference in recorded strain before and after application of the stress. Figure 1.2(a) shows what might be observed at the output of the amplifier in this situation. Here there is a moderately high level of white noise superimposed upon the signal component.

In due course it is shown that the effect of the filter is to produce the output voltage v_0 which is the 'running average' of the input voltage v_a. The function is illustrated in figure 1.2(b). Here, at time t, v_0 is the average of v_a over the period shown from $t - T_r$ to t. Thus T_r is termed the 'averaging time' for the filter.

In order to construct the output function v_0, the averaging box in figure 1.2(b) must be made to 'run' along the input function v_a in (a), thereby describing the output function v_0 in (c). Clearly the effect is to reduce the noise. It is shown subsequently that the amplitude $\tilde{v}_n$ of the noise component of v_0 is proportional to $T_r^{-1/2}$, i.e. $\tilde{v}_n \propto T_r^{-1/2}$.

The other feature that is clear from figure 1.2 is that it takes the finite time T_r

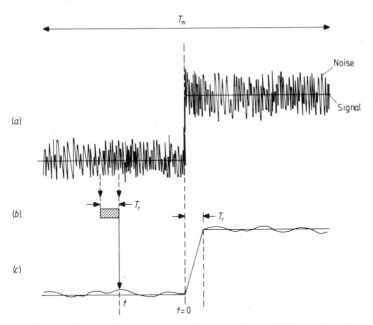

Figure 1.2 Diagram showing how the low-pass filter of figure 1.1 reduces the white noise amplitude by forming a running average of the filter input. (*a*) Filter input (v_a in figure 1.1), (*b*) running average process, (*c*) filter output v_o.

for v_o to respond to the step in v_{in}. Thus T_r is also known as the filter 'response time'.

Visual averaging

The second method of signal recovery to be considered is visual averaging. Looking at the filtered noisy signal step in figure 1.2, the basic requirement is to determine the difference between the signal before and after the step. Here the eye forms two averages, one before and one after the step, and subtracts one from the other. Now, because there are several noise fluctuations over each averaging period, the eye is able to determine the signal voltage to an accuracy greater than that given by the noise amplitude, even at the filter output.

The reason for this improvement is that basically it is the process of averaging the noise which reduces its effect and filter and eye are just two different methods of averaging. The reason that the noise amplitude $\tilde{v}_n$ at the filter output is proportional to $T_r^{-1/2}$ is that when white noise is averaged by any means over a period T_{av} the value of the typical noise error in the average is proportional to $T_{av}^{-1/2}$. Then, since for the filter $T_{av} = T_r$, the noise error is proportional to $T_r^{-1/2}$. When, however, the eye extends the averaging period to the period $T_m/2$ of the half step the noise error is reduced from the value of say $kT_r^{-1/2}$ given by the amplitude to $k(T_m/2)^{-1/2}$.

It is the recovery of signals from white noise by low-pass filtering and by visual averaging that constitutes the main topic of the present chapter. However, before proceeding to this detailed study we complete the overview by briefly introducing the other types of unwanted signal and signal recovery methods.

Drift and offset

The next two types of unwanted signal to be covered are offset and drift. Of these, offset is defined as that component of the total of all unwanted signal components which does not vary with time. For the resistor bridge strain gauge of figure 1.1 the two main sources of offset are the transducer (the resistor bridge) and the signal amplifier. The transducer offset arises from initial imbalance in the bridge and is easily removed by adjusting the balance resistor shown. Similarly the DC coupled signal amplifier will also be provided with a suitable balancing potentiometer. Thus, for the present example, offset is not a major problem. However, for some systems the offset cannot be so simply removed. Then an acceptable alternative is the baseline correction method to be discussed in §1.4.

Unfortunately the offset, even if initially adjusted to zero, tends to 'drift' as time proceeds. This is usually because of slow changes in the temperature of the circuit.

Now, drift error tends to increase as the time T_m spent making a measurement increases; this is directly opposed to the trend for white noise, for which the noise error varies as $T_m^{-1/2}$. Thus, drift tends to frustrate attempts to obtain a low white noise error by using a long measurement period, and we shall find that much of the strategy in signal recovery lies in devising methods of overcoming this problem.

Multiple time averaging

Two main methods are used to reduce the drift error and so allow the low white noise error associated with a long period of measurement to be realised. These are multiple time averaging (MTA) and the phase-sensitive detector (PSD) method. Of these MTA is a matter of carrying out the experiment rapidly rather than slowly, thus reducing the drift error, and then extending the averaging time to obtain the low white noise error by repeating the experiment many times and averaging the results. The use of MTA to avoid effects of drift in this way constitutes the main topic of chapter 2.

Phase-sensitive detection

Sometimes the experimental constraints do not allow the time scale of the experiment to be reduced. Here a suitable alternative is to modify the experiment in such a way that the transducer produces an AC rather than a DC signal. This allows an AC coupled amplifier to be used which does not respond to offset or drift.

Here some kind of rectifier is usually required in order to convert the AC amplifier output into a DC signal suitable for display. This, then, is the function of the phase-sensitive detector. It is the use of the PSD method to overcome drift that constitutes the main topic of chapter 3.

$1/f$ noise

The major remaining class of unwanted signal to be considered is $1/f$ noise. For both white and $1/f$ noise the names refers to the spectral distribution of the noise. For white noise the spectral density G is independent of frequency f, while for $1/f$ noise $G \propto f^{-1}$. These terms belong to the frequency domain view and will be elaborated later. For the present, the waveforms of figure 1.3(a) and (b) will suffice to identify the difference between the two. These are specimens of white and $1/f$ noise as might be observed at the output of the signal amplifier in the strain gauge of figure 1.1. Clearly the $1/f$ noise has a greater tendency to 'walk away' from the mean. This means that the advantages of using a larger measurement period T_m are less than for white noise. In fact, it will be shown that the $1/f$ noise error is independent of T_m. This contrasts with white noise, for which the error is proportional to $T_m^{-1/2}$, and drift, for which the error increases with T_m.

(a)   (b)

Figure 1.3 Types of noise seen at the output of the filter in figure 1.1: (a) white noise, (b) $1/f$ noise.

It is thus clear that $1/f$ noise too frustrates any attempt to obtain a low white noise error by increasing T_m. For low values of T_m the white noise may dominate, and then increasing T_m will cause a reduction. However, a point will be reached at which the white noise error falls below the $1/f$ noise error, which does not vary with T_m. Then any further increase in T_m will have no effect.

Fortunately, both MTA and PSD methods are effective in removing $1/f$ noise error but, in order to show how this is done, it is necessary to use the spectral view of both signal and noise. This is in contrast with the rather more direct 'time-domain' view used in the chapters listed so far.

Frequency-domain view

The frequency-domain view is a commonly adopted one and the present text is perhaps a little unusual in not adopting it from the start. However, I have preferred the directness of the time-domain approach for the initial treatment. Thus, in chapter 4 the frequency-domain or 'spectral' view will be introduced and the ground previously covered from the time-domain viewpoint re-covered from the frequency-domain viewpoint. Chapters 5 to 7 then use the frequency-domain view to show how the MTA and PSD methods remove the $1/f$ noise.

Digitisation

These seven chapters comprise the first part of the book, which deals with the recovery of a continuously varying analogue signal from the various kinds of unwanted signal. The remainder deals with one or two closely related topics. Frequently today an analogue signal must be digitised for computer storage and perhaps for further processing. In chapter 8 the problems of digitisation are covered. It transpires that the main factor of importance is that no data should be wasted.

Pulsed signals

Chapter 9 deals with pulsed signals or more general types of signal transient. Of particular interest is the case where the shape of a transient is known but the amplitude is not, being the factor to be measured. It is shown that the processing required to give minimum noise error is what is known as 'matched filtering'.

Signal timing

In chapter 10 the problem is a little different. Here again the shape of the signal transient is known but the factor to be measured is the time of occurrence, rather than the amplitude, of the signal. The necessary modifications to the matched filtering in this case are discussed.

The need to use a matched filter when determining the amplitude of a signal transient of known shape is well established. Also the modifications needed when signal timing is to be determined are reasonably well documented. However, such coverage is usually for white noise only. In instrumentation, as distinct from communications applications, it is highly likely that drift and $1/f$ noise will be the predominant unwanted signals. It is the distinctive contribution of the present text that the case of $1/f$ noise is covered also.

1.2 LOW-PASS FILTERING OF SHOT NOISE

The overview is now complete and we proceed to the main topic of the present chapter. This is the way in which low-pass filtering and visual averaging recover a required signal from white noise. There are two main types of white noise: thermal and shot noise. For all purposes of signal recovery these are indistinguishable. Thus in this section the effect of low-pass filtering on shot noise will be examined. It will be shown in particular that the amplitude $\tilde{v}_n$ of the noise at the filter output is proportional to $T_r^{-1/2}$, where T_r is the filter averaging time. This relation also holds for thermal noise.

Thermal noise

It will, nevertheless, be useful to indicate briefly the origins of thermal noise. This originates mainly from the resistors in a circuit and arises from the fact that the mobile charge carriers in a resistor are in constant thermal motion. Thus, at any one time the charge distribution over the resistor is not quite uniform. This causes a potential difference across the ends of the device. As the thermal motion proceeds, the potential fluctuates in a random manner about a mean of zero. This class of noise is discussed further in §4.8, with quantitative precision.

Shot noise

Shot noise tends to predominate in the semiconductor components in a circuit, i.e. in the diodes and the transistors. It is actually a manifestation of the fact that an electric current is not a continuum but consists of discrete current carriers, i.e. electrons and 'holes'. Thus the commonly used analogy of a stream of flowing water to represent an electric current might more properly be replaced by a stream of grains of sand. It is then the 'grainy' nature of the flow that constitutes the shot noise.

Shot noise model

Figure 1.4 gives a simple model for the shot noise in the amplifier of figure 1.1. Here it is reasonable to assume that all of the shot noise originates from the first stage, because that from later stages is subject to less gain. The transistor

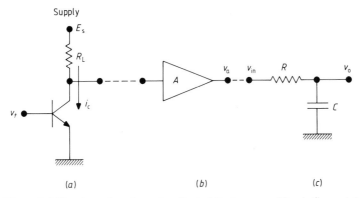

Figure 1.4 Circuit used to show the effect of the low-pass filter in figure 1.1 upon the amplifier shot noise. (*a*) First stage, (*b*) further stages, (*c*) low-pass filter.

current i_c actually consists of a random train of pulses, with each pulse corresponding to the passage of an electron from the emitter to the collector of the transistor. The area of each pulse will be q_e, the electron charge. Thus the area q of the corresponding pulse at the output of the amplifier A will be given by

$$q = q_e R_L A. \tag{1.1}$$

Figure 1.5(a) shows the sequence of these random pulses that make up the waveform for $v_a = v_{in}$ in figure 1.4. This is somewhat idealised, assuming that the amplifier contains no low-pass filter, but will do for our present purposes.

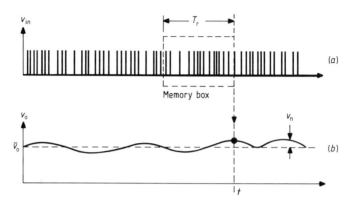

Figure 1.5 Memory box running average approximation to filtering of shot noise pulse train for transistor in figure 1.4. (a) Shot noise pulse train v_{in}, (b) running average v_o.

Running average

It will next be shown that the essential feature of the filter, in the present view, is that it forms a 'running average' of the input. Figure 1.5 illustrates this function once more. Here the output v_o at time t is the average of v_{in} over the period T_r of the memory box. Then the output function v_o is constructed as the box 'runs' along the input function. In mathematical terms

$$v_o(t) = T_r^{-1} \int_{t-T_r}^{t} v_{in}(t')\, dt'. \tag{1.2}$$

The fluctuation component v_n of v_o is the shot noise and arises because v_o is proportional to the number of pulses p in the box and p fluctuates because of the random placing of the pulses. The fluctuation time is T_r because this is the time taken for all of the pulses in the box to be replaced.

Filter response

The next step is to determine the exact relation between the input v_{in} and output v_o of the low-pass filter and thus to show that v_o is essentially the running average of v_{in}.

Consider first the 'impulse response' g_i of the filter, shown in figure 1.6(f).

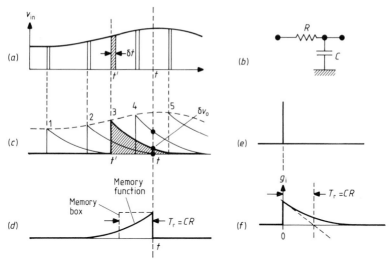

Figure 1.6 Relation between output v_o and input v_{in} for a low-pass filter. (*a*) Input v_{in}, (*b*) filter, (*c*) components of output v_o originating from elements in (*a*), (*d*) memory function, (*e*) input impulse, (*f*) impulse response.

This is the response of the filter to the unit impulse shown in figure 1.6(*e*), which is applied at time $t = 0$. Here the unit impulse is a pulse of unit area and of width approaching zero.

The input impulse leaves the capacitor with a charge. This decays exponentially through the resistor R as shown, with the decay time CR.

Next consider the shaded element at time t' of the general input waveform v_{in} shown in (*a*). For δt small, this approximates to an impulse of area $v_{in}(t')\,\delta t$. This will produce at the filter output the shaded transient shown in (*c*). At the time t, the output component δv_o will then equal the impulse area multiplied by $g_i(t - t')$, i.e. $v_{in}(t')g_i(t - t')\,\delta t$. Then summing δv_o for all values of v_{in},

$$v_o(t) = \int_{-\infty}^{t} v_{in}(t')g_i(t - t')\,\mathrm{d}t'. \qquad (1.3)$$

Looking at the transients in (*c*) following the various elements of v_{in} shown in (*a*), and numbered 1 to 5, it is clear that v_o consists of contributions from v_{in} extending back from time t at which the filter output is observed for a period roughly equal to CR. This is highly reminiscent of the running average, with the averaging period $T_r = CR$. The main difference is that the 'memory function' $g_i(t - t')$ from equation (1.3) and shown in (*d*) has an exponential 'taper'. This contrasts with the abrupt cut-off of the 'memory-box' approximation corresponding to the running average. This property is sometimes referred to as a 'fading' memory of the input.

Weighted running average

Clearly v_o is not a simple running average of v_{in} over T_r but a 'weighted' running average, the weighting given by the fading memory function $g_i(t - t')$. Actually it is only clear from equation (1.3) that v_o is a weighted *integral* of v_{in}, over the approximate period T_r. An average, whether weighted or otherwise, is said to be *properly normalised* if, when the series of values that are being averaged are all the same, the average is equal to any one of the values. But, for the filter, if v_{in} is constant, then $v_o = v_{in}$. Thus v_o is a properly normalised weighted running average of v_{in}, with the weighting given by the memory function which is the impulse response g_i reversed in time.

We now return to the main theme of the present section, which is to calculate the amplitude $\tilde{v}_n$ of the noise component v_n of the filter output v_o when the random train of impulses shown in figure 1.5(a) constitutes the input v_{in}. The approximation that v_o is the running average of v_{in} will be assumed, with T_r the averaging period. Then, with q the pulse area, v_o at any time t is given by

$$v_o(t) = p(t)q/T_r \qquad (1.4)$$

where p is the number of pulses in the box at time t.

Probability density function and standard deviation

The fluctuation in v_o arises from the fluctuation in p. To determine this it is first necessary to introduce the concepts of standard deviation and the probability density function. If the memory box is placed at a very large number of different and mutually exclusive locations on the input pulse train, a histogram of the results can be plotted as in figure 1.7, together with the overall mean $\bar{p}$. Here the horizontal coordinate is the number of times, n, that each value of p was obtained.

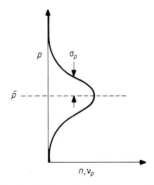

Figure 1.7 Probability density function for the number of pulses p spanned by the memory box of figure 1.5.

For present purposes it is sufficient to say that the half-width σ_p of the histogram is the 'standard deviation' and that this is given by the relation

$$\sigma_p = (\bar{p})^{1/2} \qquad (1.5)$$

provided $\bar{p} \gg 1$. Then σ_p is a good indication of the amplitude of the fluctuation in p. However, a little more precision may be preferred. Thus, if the count values n in the histogram are all divided by a common factor k to give $v_p = n/k$ and k is such that $\int v_p \, dp = 1$, the scaled histogram v_p is termed the 'probability density function' (PDF) for p.

Strictly, v_p has the 'gaussian' form

$$v_p = (\sigma_p \sqrt{2\pi})^{-1} \exp\left[-(p-\bar{p})^2/2\sigma_p^2\right] \tag{1.6}$$

and σ_p is the half-width when v_p falls from the central maximum by a factor $\exp(-1/2)$.

It will be noted that, unlike the rest of the text, the observations made in this subsection are stated rather than reasoned. They are, in fact, standard results of statistical theory and equation (1.5), in particular, is a 'foundation' statement upon which much of what follows must stand. It will therefore be of value to reflect upon its plausibility for varying values of $\bar{p}$. Notice for instance how when $\bar{p} \gg 1$ the ratio $\sigma_p/\bar{p}$ becomes small compared with 1.

From equation (1.4), the standard deviation σ_n for the noise component v_n of v_o is given by

$$\sigma_n = \sigma_p q/T_r. \tag{1.7}$$

Then, from equation (1.5),

$$\sigma_n = \bar{p}^{1/2} q/T_r. \tag{1.8}$$

Here $\bar{p} \propto T_r$. Let $\bar{r}$ be the long-term mean pulse rate. Then $\bar{p} = \bar{r}T_r$ and

$$\sigma_n = q(\bar{r}/T_r)^{1/2}. \tag{1.9}$$

This can be written as

$$\sigma_n = kT_r^{-1/2} \tag{1.10}$$

since q and r are both independent of T_r.

RMS value

The usual way to represent the amplitude of a fluctuating variable is by its 'root-mean-square' (RMS) value. For the general fluctuating variable x, the RMS value of x is $(\overline{x^2})^{1/2}$, where the average is over a period very large compared with the fluctuation time for x. The more concise symbol $\tilde{x}$ is usually adopted, i.e. $\tilde{x} \equiv (\overline{x^2})^{1/2}$.

It is clear that the RMS value $\tilde{x}$ and the standard deviation σ_x are comparable. In fact they are equal and, in this sense, the two terms are interchangeable. However, $\tilde{x}$ expresses the amplitude of the fluctuating variable, while σ_x indicates the typical deviation from the mean for just one measurement.

Since for v_n, $\tilde{v}_n = \sigma_n$, equation (1.10) becomes

$$\tilde{v}_n = kT_r^{-1/2}. \tag{1.11}$$

Thus our initial objective of showing that $\tilde{v}_n \propto T_r^{-1/2}$ has been achieved.

Experimental result

Figure 1.8(a) to (c) shows the results of a series of measurements made upon a signal step. These confirm that each time the averaging period T_r, i.e. the filter time constant CR, is increased by a factor of 10 the noise amplitude $\tilde{v}_n$ decreases by a factor of $10^{-1/2}$. It can also be seen how the filter response time and the noise fluctuation time increase in proportion to T_r. Note also how it is necessary to increase the time taken to make the measurement in order to accommodate the increase in the response time T_r.

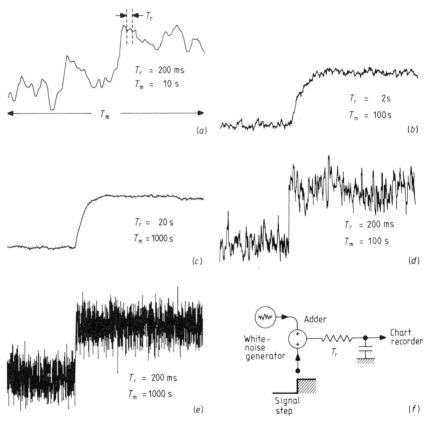

Figure 1.8 Step + white noise recorded at the output of a low-pass filter. (a)–(e) Recorded traces, (f) experimental system. In these traces, T_r is the filter response time and T_m the time taken to record the step.

Noise error in integral

Suppose that, instead of the average, the integral of the shot noise is formed over the period T_r. Since the integral is the average multiplied by T_r, and since the standard deviation σ_n of the average is proportional to $T_r^{-1/2}$, the standard deviation σ_i of the integral will be proportional to $T_r^{+1/2}$. More generally, for an integration period T_i

$$\sigma_i = k T_i^{1/2}. \tag{1.12}$$

Amplifier filter
It should, perhaps, be noted that the noise at the output of a real signal amplifier would never appear as in the idealised model of figure 1.5(a). In reality there would always be some sort of low-pass filter in the amplifier, whether by accident or design, and this would smooth the train of randomly placed impulses to produce a waveform more like that of the noise in figure 1.2(a). Here the small fluctuation time will be the response time of the filter internal to the amplifier. The external filter, with the much longer response time T_r, then extends the fluctuation time to T_r.

1.3 VISUAL AVERAGING

Visual averaging is the process whereby the averaging of a displayed noisy signal by the eye of the observer makes the overall noise error less than that given by the noise amplitude $\tilde{v}_n$. Figure 1.9 shows noise at the output of a filter of averaging time T_r superimposed upon a required signal v_s and displayed for the period T_v. If then $\tilde{v}_n = k T_r^{-1/2}$, it follows that the standard deviation for the noise averaged over the period T_v will be given by σ_v where

$$\sigma_v = k T_v^{-1/2}. \tag{1.13}$$

Then, since $T_v \gg T_r$, $\sigma_v < \tilde{v}_n$ and a reduction is obtained.

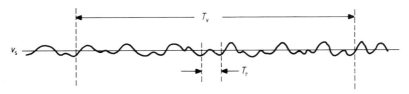

Figure 1.9 Noise at low-pass filter output superimposed on required DC signal v_s and averaged by eye over the period T_v.

Another way of looking at the process is as follows. The fluctuation time for the noise is T_r. Thus the eye averages $n = T_v/T_r$ independent fluctuations. The reduction in noise error is by $\sigma_v/\sigma_n = n^{-1/2}$. This is an example of the general truth that, if n independent random samples are averaged, the resulting noise error is reduced by $n^{-1/2}$.

Note that the final noise error σ_v is independent of the filter time constant T_r, although the noise amplitude $\tilde{v}_n \propto T_r^{-1/2}$. This is because if T_r is reduced, say, causing $\tilde{v}_n$ to rise, the number $n = T_v/T_r$ is increased, so that the eye, having more fluctuations to average, can obtain a better reduction. In fact the two effects exactly cancel, and this is obvious when it is understood that the only thing that varying T_r does is to alter the division of labour between the filter and the eye in forming the final average over the period T_v.

The relative inconsequence of the setting of T_r, regardless of its effect upon the noise amplitude, is an effect frequently not realised. In fact, T_r need only be large enough to reduce the number of fluctuations to a value small enough to allow the eye to average effectively.

Experimental result

The sequence of measurements in figure 1.8(a), (d) and (e) confirms this point. Recall that the sequence from (a) to (c) corresponds to increasing the measurement time T_m and the filter response time T_r together in decade steps. This shows the variation of $\tilde{v}_n$ with $T_r^{-1/2}$. In contrast, the sequence (a), (d), (e) increases T_m in the same way but leaves T_r unaltered. Here $\tilde{v}_n$ remains the same as for (a) but the accuracy with which the step height can be determined plainly increases, because of the increasing number of fluctuations for the eye to average. Following the above contention that the noise error is independent of T_r, we assert that, comparing say (b) and (d) (although $\tilde{v}_n$ for (b) is less than for (d)), the increased potential for visual averaging in (d) reduces the final noise error to the same as for (b). Similarly (e) should give the same final error as (c).

Resolution time

For most measurements it is possible to define a required 'resolution time' and this is the period over which visual averaging is carried out. Consider, for example, the result of figure 1.10. Here the strain gauge of figure 1.1 has been used to plot the stress–strain curve for the mechanical component. The sample component is placed in a testing machine which applies a controlled stress to the sample. The stress is increased at a constant rate (scanned). This is the independent variable. Then the resulting strain (the dependent variable) is plotted against the stress.

Here it is necessary to decide upon the number of independent points needed to define the stress–strain curve. This may be expressed as the number n_p of points or the 'fractional resolution' $\chi = n_p^{-1}$. For the present rather simple function, n_p will be fairly small. For something more complex, such as a multi-line optical spectrum, n_p would be a good deal larger.

For the stress–strain curve example, $T_{res} = T_{sc}/n_p$. The eye averages over each period of length T_{res} giving, for each time-resolved point, the noise error

$$\sigma_n = kT_{res}^{-1/2}. \tag{1.14}$$

But $T_{res} = T_{sc}/n_p = \chi T_{sc}$, so

$$\sigma_n = k(\chi T_{sc})^{-1/2}. \tag{1.15}$$

Since χ is independent of T_{sc}, this means $\sigma_n \propto T_{sc}^{-1/2}$, and so the larger T_{sc}, the better.

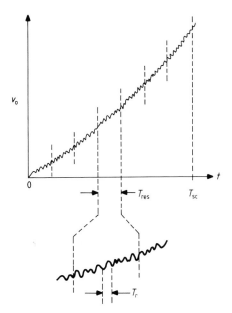

Figure 1.10 Output v_o of the strain gauge in figure 1.1 as the stress applied to the sample is scanned at a constant rate.

1.4 BASELINE SUBTRACTION

Figure 1.11 shows another method of dealing with offset. Here the example is the stress–strain curve measurement of figure 1.10 but, instead of applying the scanned stress immediately, zero stress is applied for the initial period T_b,

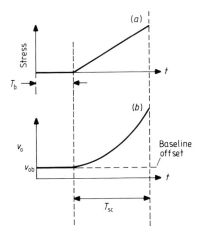

Figure 1.11 Baseline subtraction: (a) applied stress, (b) strain gauge output showing offset v_{ob}.

during which the baseline (offset) v_{ob} is measured. v_{ob} is then subtracted from each value of measured strain when the stress is applied.

T_b needs to be of non-zero length in order to allow for noise. The noise error in the baseline measurement is equal to $kT_b^{-1/2}$. Since v_{ob} is subtracted from each time-resolved value of v_o, for which the error is $kT_{res}^{-1/2}$, we require that $T_b \gg T_{res}$ if the total noise error is not to be significantly increased. Notice that the requirement is that $T_b \gg T_{res}$, not $T_b \gg T_{sc}$.

2 Multiple time averaging and drift

In the overview (§1.1) multiple time averaging (MTA) was introduced as a means of avoiding drift and so allowing the low white noise error associated with a long period of measurement to be realised. This point is covered in detail in the present chapter, but first we consider a use of MTA where the object is merely to reduce the white noise error.

Figure 2.1 shows the arrangement. Here the resistor bridge strain gauge of figure 1.1 is used to monitor the oscillatory strain in a vibrating beam after the beam has been struck by an impacting device. If the experiment is repeated a number of times and the results averaged using MTA, the white noise error is reduced as the averaging time is increased.

This kind of measurement should be contrasted with one where there is an independent variable and it is possible to vary the length of time T_{sc} used to scan through the required range of the variable. An example of this kind of measurement is the use of the strain gauge to measure the stress–strain curve of a sample, as discussed in §1.3. Here the scanned independent variable is the stress.

For such measurements, the easiest way to increase the noise averaging time, and thus reduce the white noise error, is by increasing T_{sc}. Leaving T_{sc} at the original value and repeating the scan up to the full time available has no advantage over this. However, as will be shown shortly, it is when drift (and $1/f$ noise) are present that MTA can be used to advantage with the stress–strain curve type of measurement. Here, if a short T_{sc} is used and MTA employed with a repeated scan, the low drift and $1/f$ noise associated with the short T_{sc} are combined with the low white noise error associated with a large averaging period.

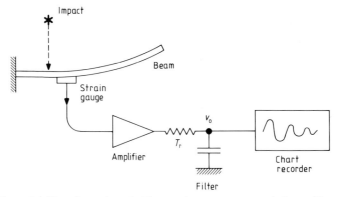

Figure 2.1 Use of a resistor bridge strain gauge to record the oscillatory response of a vibrating beam to an impact.

2.1 MULTIPLE TIME AVERAGING METHODS

Overlay averaging

There are two commonly occurring ways in which the results of a repeated measurement are averaged. The first is shown in figure 2.2(a). Here, each time the beam is struck a new noisy transient is plotted on the paper on top of the last, giving a series of 'overlaid' traces. These are then averaged by eye to give the reduced white noise error.

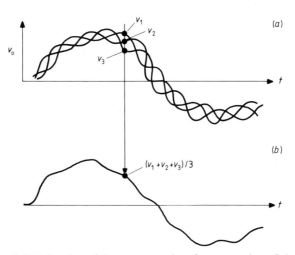

Figure 2.2(a) Overlay of three traces taken from a section of the output transient from the system of figure 2.1. (b) Computer averaging of the traces in (a).

DRIFT 19

Computer averaging

The second, and more sophisticated, method of averaging is to use a computer. Here all of the values of v_o are input to the computer, the traces are averaged by the computer and the result is displayed. For the three component traces of figure 2.2(a), this would give the result of (b).

Whichever method is used, if the number of traces averaged is n_t, the amount of time available for noise averaging is increased by the factor n_t. For one of the time-resolved values of strain on a single transient the white noise error σ_n is given by $\sigma_n = kT_{res}^{-1/2}$, where T_{res} is the resolution time. Then for n_t averaged traces, the final error for one of the time-resolved points becomes

$$\sigma_n = k(n_t T_{res})^{-1/2}. \tag{2.1}$$

Thus, the averaging of n_t traces gives an improvement of $n_t^{-1/2}$.

2.2 DRIFT

Before considering how multiple time averaging can be used to reduce drift, we first consider the subject more generally. Drift usually arises mainly from changes in the ambient temperature, draughts, etc. These influence the offset at the signal amplifier input. The system enclosure tends to act as a low-pass filter with a time constant T_d of a minute or so. Thus the fluctuation time has this value also.

Sloping baseline correction

Figure 2.3 shows an example of such drift, as seen at the output of the strain gauge. For an experiment such as the stress–strain curve measurement, if the stress scan time T_{sc} is somewhat less than the drift fluctuation time T_d as shown, then the drift approximates reasonably well to a constant rate ramp over the scan period.

When the drift rate is truly constant, it is possible to correct for the drift exactly, using the method of figure 2.4. This is similar to the simple baseline correction system of figure 1.11, but now the baseline is measured both before

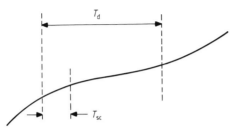

Figure 2.3 Sample of drift of fluctuation time $T_d \gg$ scan time T_{sc} for stress–strain curve measurement.

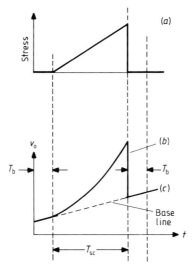

Figure 2.4 Sloping baseline correction method for drift of constant rate, as applied to stress–strain curve measurement. (*a*) Applied stress, (*b*) measured strain + drift, (*c*) reconstructed baseline.

and after the scan, in each case with the applied stress zero. From the measured values, the entire sloping baseline can be reconstructed. Then the appropriate baseline value can be subtracted from each recorded value of measured strain during the scan.

As before, the finite period T_b is needed for the baseline measurement. This is to reduce white noise error. In order for the noise error in the baseline value to be small compared with that in the measured strain, we again require that $T_b \gg T_{res}$, rather than $T_b \gg T_{sc}$.

In reality the drift rate is not constant. Figure 2.5 shows the error that results in the stress–strain curve measurement for both simple and sloping baseline

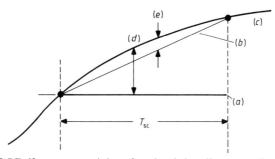

Figure 2.5 Drift errors remaining after simple baseline correction of figure 1.11 and sloping baseline correction of figure 2.4. (*a*) Simple baseline, (*b*) sloping baseline, (*c*) actual drift, (*d*) error for (*a*), (*e*) error for (*b*).

correction. Clearly, provided $T_{sc} \ll T_d$, the second method still gives a reduced drift error, although not now zero.

Drift reduction by multiple time averaging

It is clear that for either of the baseline correction methods of figure 2.5 the drift error can be reduced by reducing T_{sc}. With $T_{sc} \ll T_d$ as shown, the simple correction gives a drift error approximately proportional to T_{sc} and the sloping baseline correction method gives a decrease with T_{sc} which is even more rapid. Unfortunately, white noise error is proportional to $T_{sc}^{-1/2}$, so reducing T_{sc} in this way gives an increased noise error. This is avoided by repeating the scan over and over in the time available and averaging the results. In this way a large measurement period can be used to give the same low white noise error as for one long scan taking the entire period, but with the actual scan time reduced to the degree necessary to make the drift error negligible.

Here it is assumed that, when a short scan is repeated and the results averaged, the final drift error is the same as for the original short scan. It is conceivable that the effect of repetition and averaging would be an increase in drift error with the number n_t of scans averaged. Figure 2.6 shows that this is not so. This shows the drift error for a series of scans, using simple baseline correction. Here the overall period $n_t T_{sc}$ is somewhat less than the drift fluctuation time T_d. When $n_t T_{sc} \ll T_d$ the drift rate is nearly constant over the overall period $n_t T_{sc}$, and then the averaged drift error is the same as for one of the short component scans. When, on the other hand, $n_t T_{sc} > T_d$ but still with $T_{sc} \ll T_d$, the drift error will reverse sign and the averaged drift error will actually be less than the low value for one of the component scans.

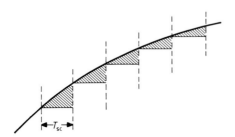

Figure 2.6 Drift errors for simple baseline correction with repeated scan.

The same is true when sloping baseline correction is used, although there, for the relative values of T_{sc} and T_d shown, the drift error is so small as to be unrepresentable on the diagram.

It is also possible to use multiple time averaging without either type of baseline correction. Then the 'tilt' component of the averaged drift error is the same as when simple baseline correction is used. However, the major part of the drift is then converted into a further component of offset. This is added to the original offset, which is now also present. The additional offset component does increase with the number n_t of scans averaged and, for constant rate drift, is proportional to n_t.

For some measurements, offset is of no importance. For example, for the measurement of a signal step the offset can simply be ignored. For this class of measurement therefore the baseline correction is not strictly required. This contrasts with the above stress–strain curve measurement where offset error is important. Here, for multiple time averaging to be effective, some type of baseline correction is required.

2.3 MULTIPLE TIME AVERAGING BY OSCILLOSCOPE

It is important to realise, when predicting the noise error of a measurement, that the normal oscilloscope display may incorporate a high degree of overlay multiple time averaging of the type shown in figure 2.2(a). Here a number n_t of traces are overlaid, where $n_t = T_{me}/T_t$ and where T_t is the oscilloscope time-base period and T_{me} is a 'memory time' compounded from the CRT screen persistence, the persistence of the observer's eye and the observer's visual memory. The noise error is thus reduced from the value $kT_{res}^{-1/2}$ given by equation (1.14) by the factor $(T_{me}/T_t)^{-1/2}$ to give

$$\sigma_n = k(T_{res}T_{me}/T_t)^{-1/2}. \tag{2.2}$$

Clearly, for very large values of $n_t = T_{me}/T_t$ this is somewhat degraded, as the eye becomes 'saturated'.

2.4 MULTIPLE TIME AVERAGING BY COMPUTER

Hardware and program

Figure 2.7 indicates how a computer is used for multiple time averaging. Here the signal averaged is that from the vibrating beam of figure 2.1. The beam is struck periodically using the impact transducer which is drawn by the pulse generator output (A). Here the pulse period T_p is made somewhat larger than the decay time for the transient oscillators of the beam, which is shown in (B).

The overall program MTAV for averaging and display is summarised in figure 2.8 and includes the procedures ADNT, ADOT and DISP (figures 2.9–2.11).

The signal transient is sampled every T_s seconds, as shown in (B), giving a total of $n_s = 500$ samples between each impact. Within the computer is a data buffer of n_s locations which is initially cleared. This is the first step of MTAV. Then, following each impact, the n_s samples are added to the value stored at each successive location in the data buffer. For one transient this is detailed in the procedure ADOT. This is repeated for the required number n_t of transients as indicated in ADNT, which is the second step of MTAV.

Then the data buffer contents represent the sum of n_t transients. Thus, to

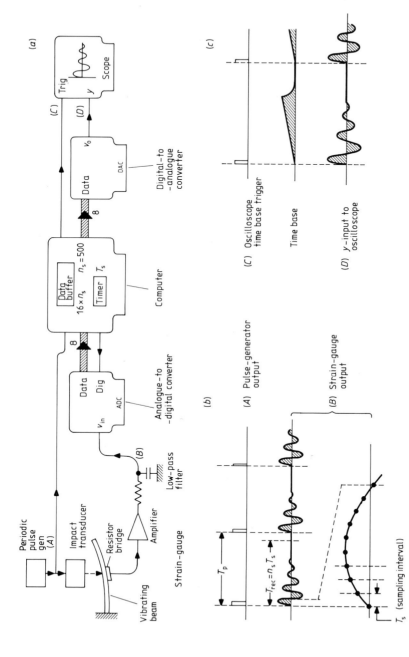

Figure 2.7 Multiple time averaging by computer, as applied to vibrating beam measurement of figure 2.1. (a) System diagram, (b) waveforms for logging, (c) waveforms for display.

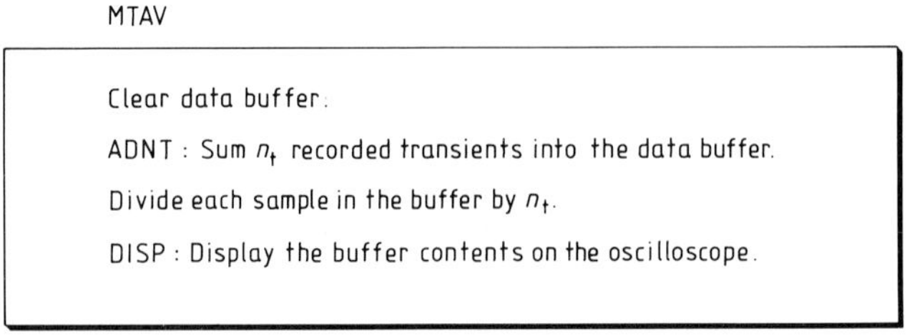

MTAV

> Clear data buffer.
>
> ADNT : Sum n_t recorded transients into the data buffer.
>
> Divide each sample in the buffer by n_t.
>
> DISP : Display the buffer contents on the oscilloscope.

Figure 2.8 Structure of MTAV—main program for multiple time averaging, using the system of figure 2.7.

ADNT

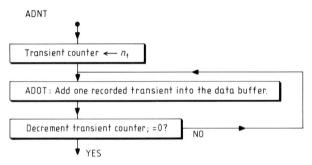

Figure 2.9 Procedure ADNT sums n_t recorded transients into the data buffer.

DISP

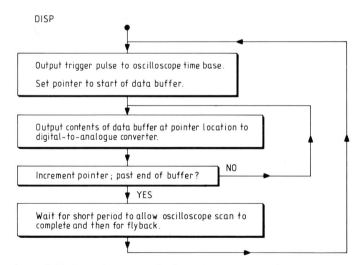

Figure 2.10 Procedure DISP displays the contents of the data buffer continuously on the oscilloscope screen.

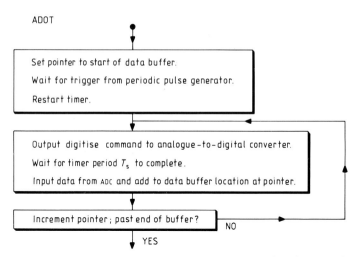

Figure 2.11 Procedure ADOT adds one recorded transient into the data buffer.

form the average each value must be divided by n_t. This is the third step of MTAV.

At this stage the contents of the data buffer represent the averaged transient of the type shown in figure 2.2(b) and it merely remains to display this as in DISP. Here, a trigger pulse (C) is output by the computer to trigger the oscilloscope time-base and then the data buffer contents are output sequentially to the digital-to-analogue converter. This sequence is repeated indefinitely, allowing a brief interval between the display scans for the time-base flyback to take place.

Notice that the number of bits for each location in the data buffer is 16, while the ADC is only an 8-bit device. Generally, if i is the number of bits for the ADC and 2^j is the number of transients to be averaged, then the number of bits k required to store the sum will be $i+j$. For any lower value of k, overflow is possible.

For the present example, $k=16$ and $i=8$, so that the number of transients that can be averaged is limited to $2^8 \equiv 256$. For a larger number of transients, the number of bits per location in the data buffer will need to be increased from the present value of 16.

The only other requirement is that the conversion time for the ADC must be less than T_s.

Continuous display

An unfortunate aspect of the above arrangement is that the progress of the summed transient is not monitored as it builds up in the data buffer. This can waste a good deal of time because, if anything is wrong, calling for some adjustment and a restart, the operator is not made aware of this until the full sequence of n_t transients has been accumulated.

One way to obtain continuous monitoring is to run the display routine interleaved with the logging routine, on some sort of interrupt basis. Then for the present arrangement the eight most significant bits from the 16-bit data buffer word are output to the DAC for display. This has the effect of dividing the displayed value by 2^8, the present value of n_t. Then, as the sequence of 2^8 transients builds up in the buffer, the displayed transient is seen to grow from zero to the full range of the DAC. Because the signal amplitude increases directly with the number of transients n_t summed to date, while the noise amplitude increases with $n_t^{1/2}$, the signal appears to grow out of the less rapidly increasing noise.

The above statement concerning the increase in noise should perhaps be justified. In §2.1 it is shown that, when n_t transients are averaged, the standard deviation for the averaged white noise is reduced by $n_t^{-1/2}$. Thus, for the sum formed prior to averaging, the noise amplitude will be proportional to $n_t^{1/2}$.

A further advantage of continuous monitoring is that if, during a successful run, it is observed that an adequately noise-free signal has been accumulated before the end of the sequence of n_t transients, it is possible to save time by stopping the logging sequence early. At this stage the usual display-only mode will be entered, and it may then be necessary to increase the effective gain of the display, because fewer than the usual number of transients will have been accumulated. The simplest method is to output to the DAC, not the most significant eight bits of the 16-bit data buffer word, but a suitable sequence of bits of lower significance.

Continuous time averaging

With or without continuous display, the above method of time averaging still has the property that averaging is carried out over a given, usually predetermined, period and then has to stop, for fear of overflowing the data buffer. In some instances a mode of averaging more akin to the running average formed by a low-pass filter is more suitable. Here the displayed transient would represent the average of the last n_t transients to be recorded. This would have the dual advantage of giving a continuously updated account of the signal transient, and also of not overflowing the data buffer.

The low-pass filter is not suitable for multiple time averaging in its normal form, because the values that it averages are consecutive in time. For multiple time averaging, in contrast, the averaged values are taken from points widely separated in time, with the values for any one data buffer location all taken from different input transients.

If, however, a method can be found for making the computer perform the basic function of the low-pass filter, then the inherent flexibility of the computer should make it an easy matter to adapt the process to the requirements of multiple time averaging.

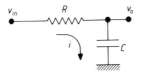

Figure 2.12 Simple low-pass filter to be simulated by computer.

We therefore start by determining how the computer can be made to carry out the function of the low-pass filter shown in figure 2.12. For the filter shown, $dv_o/dt = i/C$ and $i = (v_{in} - v_o)/R$, so that

$$dv_o/dt = (v_{in} - v_o)/T_r \qquad (2.3)$$

where $T_r = CR$.

Thus, if the initial value of v_o is known, the subsequent values are determined by the input function v_{in} and equation (2.3).

For the computer to carry out this function, the simple CR circuit must be replaced by the ADC, the computer, and a DAC. The ADC converts the analogue input voltage v_{in} to digital form, the computer executes the process of equation (2.3) and the DAC converts the result into the output voltage v_o.

Suppose that v_{in} is digitised at regular brief intervals of length Δt. Then, to reflect the process of equation (2.3), the digital number representing v_o must increase at each interval by the amount Δv_o given by

$$\Delta v_o/\Delta t = (v_{in} - v_o)/T_r. \qquad (2.4)$$

This implies the algorithm

$$v_o \leftarrow v_o + (v_{in} - v_o)\, \Delta t/T_r. \qquad (2.5)$$

Here T_r is the period over which both CR filter and computer average v_{in}. For the computer, this will equal $n_t\, \Delta t$, where n_t is the number of digitised input values averaged. Thus the algorithm can be expressed in the form

$$v_o \leftarrow v_o + (v_{in} - v_o)/n_t. \qquad (2.6)$$

Adaptation to multiple time averaging is simple; all that is necessary is to replace the simple summing algorithm in the present program by that of equation (2.6). Here, v_{in} represents the value output by the ADC, and v_o the current value of the corresponding word in the data buffer.

For the present arrangement the most suitable way of executing the continuously averaging algorithm would be as follows. First the 8-bit input representing v_{in} is regarded as the 8 most significant bits of a 16-bit word, for which the remaining 8 bits are zero. Then the stored 16-bit value representing v_o is subtracted from this value. Next the division by n_t, here 2^8, is performed by shifting the 16-bit difference down by eight bits. Then the divided difference is added to the stored value for v_o.

If this method is used, and the eight most significant bits of the data buffer are output to the DAC as before, then the initial growth of the displayed signal will be the same as for the simple summing algorithm. However, when the 16-bit numbers representing the values v_o of the stored transient begin to approach the effective 16-bit numbers representing the input transient v_{in}, the rate of growth decreases. The rise then becomes exponential and eventually terminates when the magnitude of the stored transient becomes equal to that of the input transient. Thereafter the signal transient becomes largely stable, reflecting only changes that are sustained for a period comparable with the time taken to record n_t transients, as for example might occur if the amplitude of the signal transient was gradually increasing due to some deterioration in the structure of the vibrating beam.

Throughout, the noise amplitude will remain at the reduced level corresponding to the averaging of n_t samples.

3 Phase-sensitive detector methods

It has been stated previously that the phase-sensitive detector (PSD) method is a useful alternative to multiple time averaging (MTA) for avoiding errors due to drift and $1/f$ noise, and thus that either can be used to prevent these effects from frustrating the low white noise error obtainable by using a long period of measurement. The principal object of the last chapter was to show that MTA can be used to reject drift in this way. In this chapter the drift rejecting properties of the PSD system are explained. The discussion of the ability of either system to reject $1/f$ noise is considered later, for the PSD in chapter 7.

The present chapter ends with a number of miscellaneous applications of the PSD method.

3.1 ADAPTATION OF RESISTOR BRIDGE STRAIN GAUGE

The example used to introduce the PSD method is the usual one of the resistor bridge strain gauge of figure 1.1, and figure 3.1 shows this adapted to the PSD method. Here the DC supply to the bridge is replaced by the AC supply e_g and transformer shown in figure 3.1. Figures 3.1(b), (c) and (e) then show how the strain s is converted to the AC signal v_a at the amplifier output.

The phase-sensitive detector then demodulates the signal to give the voltage v_p in figure 3.1(f). The low-pass filter then extracts the slowly varying mean, which is displayed as v_o. The filter operates by forming the running average of v_p over the period T_r of the filter and this requires that $T_r \gg f_0^{-1}$, where f_0 is the signal frequency.

Notice that $v_o \propto s$, the strain. Were the simple full-wave rectifier of figure 3.2 to be used in place of the PSD, the output would be as in figure 3.1(g) and v_o would become a distorted version of s.

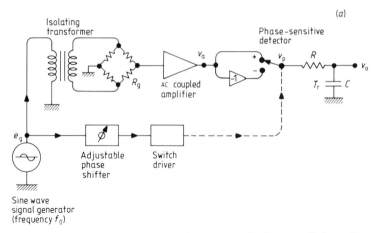

Figure 3.1(*a*) The phase-sensitive detector method as applied to the resistor bridge strain gauge of figure 1.1: system diagram.

The phase shifter in figure 3.1 is needed to ensure that the switching of the PSD is correctly phased, as in figure 3.1(*d*). This is to correct for phase errors in the transformer, etc, which, in some cases, are so small as to make the phase shifter unnecessary.

It is actually unlikely that the PSD switch will be mechanical as shown, although a relay is sometimes used. The use of some sort of transistor switch is much more likely.

3.2 AC COUPLED AMPLIFIER

There are two ways in which the PSD system rejects drift: one is that an AC coupled amplifier can be used in place of the usual DC coupled signal amplifier and the other is that the PSD itself tends to reject drift. In this section the drift-rejecting properties of the AC coupled amplifier are considered.

Figure 3.1 shows the AC coupled amplifier replacing the DC coupled amplifier of figure 1.1. Figure 3.3 shows the essential features of the AC coupled amplifier. Here the single-stage transistor amplifiers are separated by simple high-pass *CR* filter sections, and the arrangement is terminated with a unity-gain isolator to drive the PSD. The exact arrangement would now be a little out of date and a feedback amplifier section, such as that in figure 3.3(*b*), would probably replace each single stage of figure 3.3(*a*). However, the response is the same, and for our present purpose it is convenient to analyse system (*a*).

Offset
The first step is to show that the AC coupled amplifier rejects offset. One way of showing this is to consider the response of one of the *CR* networks to a step

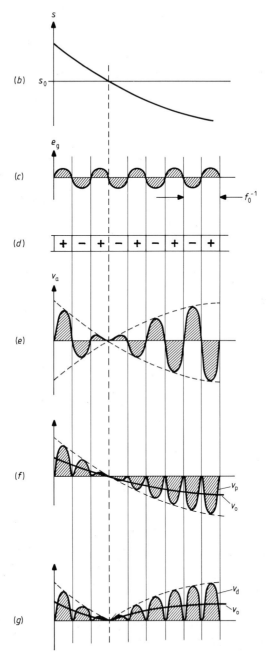

Figure 3.1 (*continued*) The phase-sensitive detector method as applied to the resistor bridge strain gauge. (*b*) Strain, (*c*) signal generator, (*d*) switch state, (*e*) amplifier output, (*f*) PSD and filter outputs, (*g*) detector and filter outputs when PSD is replaced by full-wave rectifier of figure 3.2.

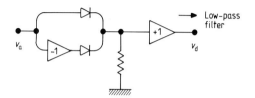

v_a v_d Low-pass filter

Figure 3.2 Full-wave rectifier.

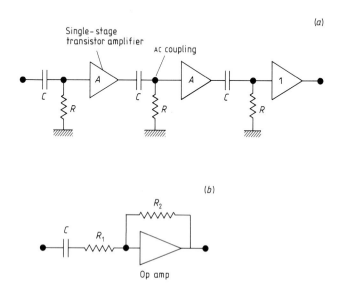

(a)

Single-stage transistor amplifier   AC coupling

(b)

R_2

C R_1

Op amp

Figure 3.3 AC coupled amplifier circuits. (a) Cascaded single-stage (time constant $= CR$) transistor amplifiers, (b) feedback amplifier (gain $= R_2/R_1$, time constant $= CR_1$).

applied to the input. Figure 3.4 shows the step (b) applied to the network (a) giving the response (c). Here the output responds initially in full to the input change, but then decays to zero as the capacitor C charges through the resistor R. Thus, after the initial transient, the response to DC applied to the input is zero.

Square wave signal response

It is important to show that the AC coupling networks do not attenuate the required signal. Depending upon the way in which the system is adapted, this can be a square wave or a sine wave, simply depending upon whether the generator e_g in figure 3.1 is a square or a sine wave generator. Figure 3.5 shows the effect when a square-wave signal is applied to one of the CR sections of figure 3.3. It is clear that, to avoid attenuation, we require that $T_r \gg T_0$, where T_0 is the signal period and T_r is the filter time constant CR.

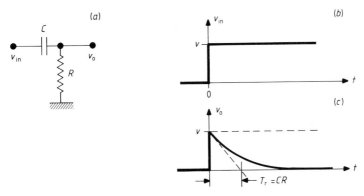

Figure 3.4 Step response of AC coupling element in figure 3.3(a). (a) AC coupling network, (b) input step, (c) output response $v_o = V \exp(-t/T_r)$.

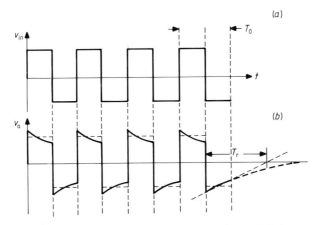

Figure 3.5 Response of AC coupling element in figure 3.3(a) to a square wave of period T_0 somewhat less than the filter response time T_r. (a) Input, (b) output.

Drift response

In this chapter we shall consider mainly the ability of the AC coupled amplifier to reject drift of constant rate. Drift of varying rate requires the spectral view of the next chapter and is considered later, in chapter 5.

To demonstrate the constant-rate drift response, we first consider the response of one CR section to the input signal of figure 3.6(a). This is a constant-rate ramp starting abruptly at time $t=0$. Consider first the 'steady-state' response at a time well after the initial transient. Here the output v_o is constant. Thus, for the capacitor voltage v_c, dv_c/dt will equal dv_{in}/dt. But the capacitor current $i = C \, dv_c/dt$, so $i = C \, dv_{in}/dt$. Since $v_o = iR$, one can write $v_o = CR \, dv_{in}/dt$. But dv_{in}/dt has the constant value $\dot{v}_{in}$, so that v_o has the constant value $CR\dot{v}_{in}$ as shown.

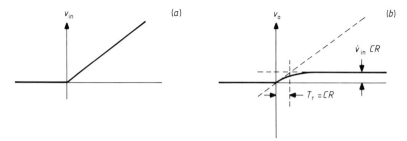

Figure 3.6 Response of AC coupling element of figure 3.3(a) to a ramp input. (a) Input, (b) output.

It appears then that a single high-pass filter section does not entirely reject drift. The drift is converted to a steady output. This, of course, is easily eliminated by adding another high-pass filter section. This is why a typical AC coupled amplifier will have at least two high-pass sections; the arrangement of figure 3.3(a) has three.

3.3 DRIFT REJECTION OF PSD

Although the AC coupled amplifier commonly used with the PSD system has drift rejecting properties, as explained above, the use of the AC coupled amplifier is not actually fundamental to the technique, because the PSD itself tends to reject drift, and offset, and indeed $1/f$ noise. It is the purpose of this section to show how the PSD rejects offset and drift.

Figure 3.7 illustrates the above point. Here v_a represents drift and offset at the amplifier output, i.e. the PSD input. This slowly varying signal is converted to a square wave (c) with a corresponding slow variation in amplitude. Then, because the frequency of the square wave is the signal frequency f_0, and because $f_0 \gg f_r$, where f_r is the cut-off frequency of the low-pass filter that follows the PSD, the filter virtually rejects the square wave, leaving only the small residual component (d).

PSD imperfections

There are two kinds of imperfection in a real PSD circuit which impair its ability to reject drift and offset, and therefore make it desirable to retain the AC coupled amplifier. The first is that the magnitudes of the gain in the forward and reversed states of the switch may not be exactly equal. This causes a small DC component at the PSD output which, unlike the output square wave component resulting from the offset, will be accepted by the low-pass filter. Then, naturally, if the input offset component drifts, so also will the small DC component at the output.

The other potential imperfection of the real PSD circuit is that, in general, the

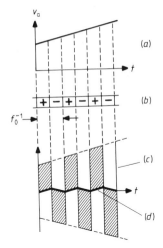

Figure 3.7 Waveforms showing how the PSD circuit of figure 3.1 avoids the effects of drift and offset. (a) Drift and offset at amplifier output, (b) PSD switch states, (c) PSD output v_p, (d) low-pass filter output v_o.

periods for the forward and reversed states of the switch will not be exactly the same. This will have essentially the same effect as when the gains for the two states are unequal.

3.4 WHITE NOISE ERROR

It has now been shown that when a system is adapted to phase-sensitive detection, drift is removed. It remains to be shown that the adaptation gives no increase in white noise error.

Figure 3.8 allows this point to be demonstrated. (a) shows the essentials of the resistor bridge strain gauge of figure 1.1 and (b) shows an adaptation to phase-sensitive detection. Here the sine-wave generator e_g of figure 3.1 is replaced by the DC source E and the reversing switch of figure 3.8(b). This constitutes a square-wave generator and allows a fair comparison of (a) and (b), because the same primary power source E is used in both cases.

The first point to note is that the signal output for the two systems, v_a in (a) and v_p in (b) are the same. This is because the two reversing switches in (b) cancel each other. The waveforms in (c)–(f) confirm the point. In (c) there is an output component v_{aa}. This is the signal component resulting from the strain which unbalances the resistor bridge. (d)–(f) then show that the PSD system output v_p is the same, i.e. $v_p = v_{aa}$.

Waveforms (g) to (j) now show the result of adding noise. Here the fluctuation time is determined by the response time of the signal amplifier. As before, the signal components v_{aa} are the same at each output. The noise,

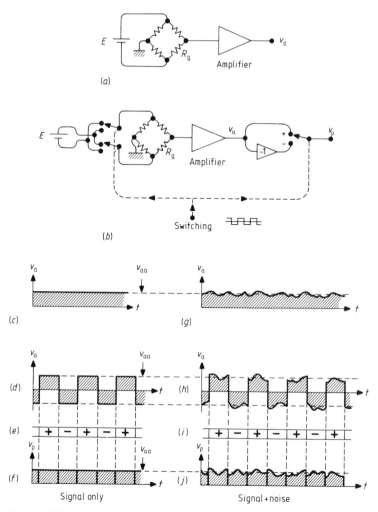

Figure 3.8 Diagram showing how adapting the resistor bridge strain gauge to the PSD method leaves the system response to both signal and white noise unaltered. (*a*) Simple resistor bridge strain gauge, (*b*) adaptation to PSD. (*c*) and (*g*), amplifier output for (*a*); (*d*) and (*h*), amplifier output for (*b*), (*e*), (*i*) PSD switch states, (*f*), (*j*) PSD output.

however, differs slightly. That at the PSD output is subject to periodic reversals, due to the PSD reversing switch. For either system the output noise is passed through the final low-pass filter of time constant T_r (not shown). It is fairly clear that the noise amplitude after this final filtering will not be influenced by these periodic reversals, and so the final noise amplitude is the same for both systems. This, in fact, is so and is argued more rigorously from the frequency-domain viewpoint in chapter 7.

It has now been established that both multiple time averaging and the phase-sensitive detector method are effective in removing drift and offset without increasing white noise error. Thus either can be used, in principle, to allow the low white noise error associated with a large measurement time to be realised. In practice, however, the choice between the two may be limited. For example, to use MTA to reduce drift when measuring a stress–strain curve requires making the stress scan time T_{sc} small. In practice a sufficiently low value of T_{sc} may not be possible.

3.5 MISCELLANEOUS APPLICATIONS

The arrangement of figure 3.1 is not the total solution to drift and offset for the resistor bridge strain gauge. While amplifier drift and offset are eliminated, transducer drift and offset remain: if the bridge is initially unbalanced this constitutes transducer offset; and when the imbalance changes—say due to thermal changes in the relative values of the bridge resistors—this constitutes transducer drift.

A possible, if far-fetched, solution can be found for the stress–strain curve measurement. Here the original DC supply is used to drive the resistor bridge, instead of the AC supply of figure 3.1. Also, instead of scanning the applied stress uniformly as in figure 3.9(a), the stepped scan of (b) is used. This produces an AC signal component of the frequency f_0 shown, which is then processed by the AC coupled amplifier and PSD in the usual way. For this arrangement, in contrast with that of figure 3.1, the transducer drift and offset is not modulated to produce an AC signal component, and thus is rejected.

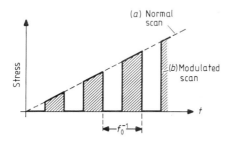

Figure 3.9 Stress modulation to avoid transducer offset and drift in stress–strain curve measurement. (a) Normal scan, (b) modulated scan.

For this particular application, the idea is impractical, because the inertia of the testing machine would severely restrict the value of f_0. However, the idea is sound in principle and serves to illustrate just one of the great variety of ways in which the PSD method can be refined to allow rejection of various kinds of unwanted response.

It is not really possible to lay down general principles beyond this point. Each system will present its own problems, and it is a matter for the ingenuity of the experimenter to devise a method which will modulate the variable that it is wished to measure, while leaving unmodulated all unwanted effects. Thus we finish by giving a few rather more practical examples of the kind of strategy that is adopted.

Choppers
Figure 3.10 shows one of the simplest methods for converting a DC signal into an AC signal suitable for AC coupled amplification and phase-sensitive detection. Here the vibrating switch contacts 'chop' the signal to produce a square wave of amplitude equal to the initial slowly varying DC voltage.

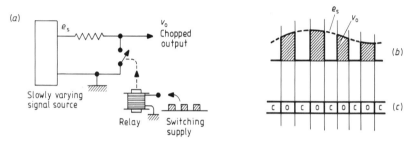

Figure 3.10 Relay-operated chopping circuit. (a) Circuit, (b) signals, (c) switch states (o = open, c = closed).

Disadvantages of the arrangement are the obvious problems of wear associated with an electromechanical system, and also the thermoelectric EMF that is generated across the contacts, even in the absence of a signal.

Both of these problems are eliminated if the relay is replaced by a field-effect transistor switch. Then the main limitation is the tendency for the switching voltage applied to the gate of the transistor to be coupled through to appear at the output, either via leakage resistance or junction capacity.

Optical chopping
Figure 3.11 shows how the PSD method can be used to remove drift and offset in an optical emission spectrometer. Here the optical source shown emits light exhibiting a series of spectral lines. To record the spectrum, a scanning monochromator is placed between the source and broad-band photodetector. The monochromator is an optical filter designed to transmit light of, or in the close region of, one wavelength only. The value of this wavelength can be varied and here is linearly scanned to plot the spectrum.

If now the light beam is chopped, using the toothed wheel shown, the photodetector output v_d becomes modulated as shown in (b). Then the AC coupled amplifier accepts the alternating component of v_d (while rejecting the

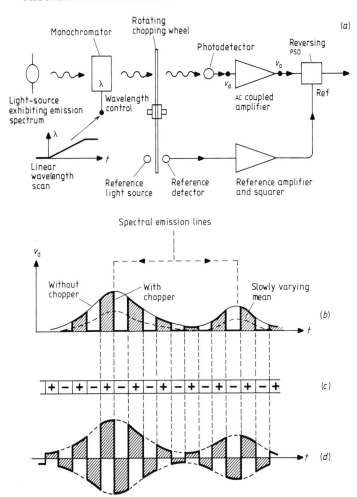

Figure 3.11 Use of the PSD method in optical emission spectroscopy. (a) Scheme of experiment, (b) photodetector output, (c) PSD reference, (d) output of AC coupled amplifier.

slowly varying component), to give the output waveform shown in (d). This is then processed by the PSD in the usual manner.

The arrangement is perfectly suitable for measuring weak emission spectra, avoiding the effects of drift, both in the amplifier and the 'transducer', here the photodetector. Another problem arises, however, when measuring weak absorption spectra. Figure 3.12 shows in this case what happens when the intensity of the, now broad-band, light source drifts. The drift, in spite of the optical chopping, is converted to appear ultimately at the PSD output.

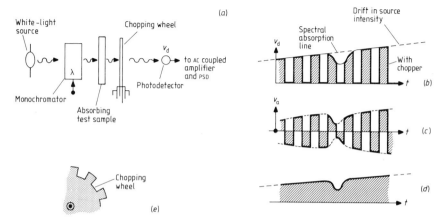

Figure 3.12 Use of chopping wheel in absorption spectroscopy, showing how drift in intensity of light source remains. (*a*) System diagram, (*b*) photodetector output, (*c*) output of AC coupled amplifier, (*d*) output of low-pass filter following PSD, (*e*) chopping wheel.

Derivative modulation

Figure 3.13 shows a method of overcoming the above problem. Here the method of modulating the signal to produce an AC component is not to chop the light beam, but to modulate the independent variable, here the mono-chromator wavelength λ. Figure 3.13(*a*) and (*c*) show how the modulation is applied. Here it is assumed, for simplicity, that λ is electrically controllable, so that λ is proportional to the control voltage v_λ. Then (*b*)–(*d*) show how the modulation causes the AC component to be superimposed upon the normal signal peak seen at the output of the photodetector. It is clear that, using this technique, drift in the intensity of the light source is not converted into an AC signal.

To complete the sequence, figure 3.13(*e*) shows the AC component of the signal after amplification by the AC coupled signal amplifier. Then (*e*)–(*h*) show how, as usual, the PSD and associated low-pass filter convert the AC signal back to DC in a form suitable for display on a chart recorder. Figure 3.14 may help to clarify the relation between figures 3.13(*b*)–(*d*) by showing the position for certain fixed values of the linear scan component of v_λ. The diagram shows clearly how the phase of the AC component reverses as the slope of the signal peak changes sign, thus requiring the final detector to be phase sensitive.

The technique is referred to as 'derivative modulation' because the output, v_o in figure 3.13(*h*), represents the first derivative of the original signal peak.

A potential disadvantage of the arrangement is that, in order to avoid distortion, the amplitude of the wavelength modulation has to be made small compared with the width of the spectral line. This causes the signal amplitude to be proportionally less than that of the original signal peak at the

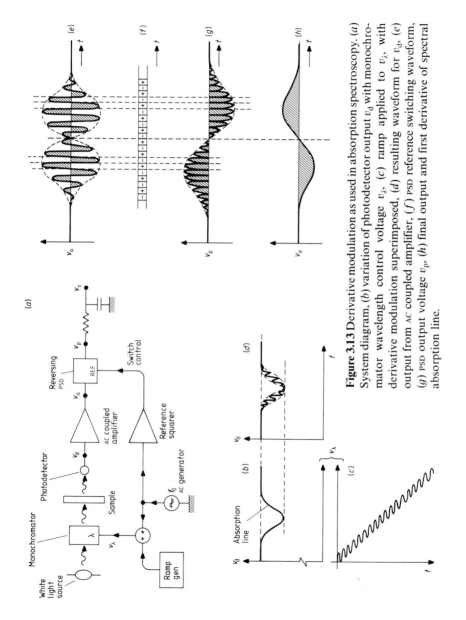

Figure 3.13 Derivative modulation as used in absorption spectroscopy. (*a*) System diagram, (*b*) variation of photodetector output v_d with monochromator wavelength control voltage v_λ, (*c*) ramp applied to v_λ, with derivative modulation superimposed, (*d*) resulting waveform for v_d, (*e*) output from AC coupled amplifier, (*f*) PSD reference switching waveform, (*g*) PSD output voltage v_p, (*h*) final output and first derivative of spectral absorption line.

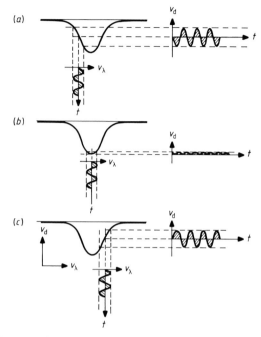

Figure 3.14 Relation between curves (b)–(d) of figure 3.13 for three fixed values of the ramp generator output, with the modulation applied.

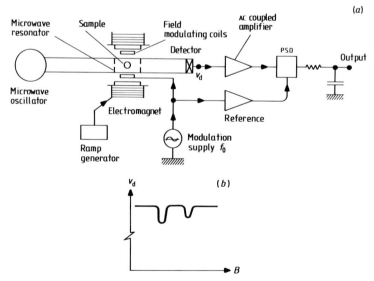

Figure 3.15(a) PSD method applied to magnetic resonance spectroscopy. (b) Two absorption lines in the magnetic resonance spectrum $(B =$ magnetic field strength).

photodetector output. Since the noise level remains the same, this represents a loss in signal-to-noise ratio and thus a compromise has to be reached. For weak signals the modulation amplitude would be made approximately equal to the line width. This will give the maximum signal amplitude possible, which is almost equal to the height of the original signal peak. Thus, there is no significant loss in signal-to-noise ratio, but some distortion has to be accepted. Then for stronger signals the modulation would be somewhat reduced, removing the distortion but somewhat degrading the signal-to-noise ratio.

Magnetic resonance spectrometer

The system of figure 3.13 would probably not be used in practice, because the mechanical inertia of the monochromator would make it difficult to use a frequency of modulation high enough to avoid $1/f$ noise. The reason for using the optical spectrometer as the example with which to introduce the technique of derivative modulation was to allow direct comparison with the optical chopping system.

A more likely application of the technique would be to magnetic resonance spectroscopy, as shown in figure 3.15. Here the extent to which the sample absorbs microwave energy of a fixed frequency varies with the strength of applied magnetic field B as shown in (b). Thus B in the magnetic resonance spectrometer assumes the significance of v_λ in the optical spectrometer and all else is the same.

4 Spectral view of signal recovery

The main object of this chapter is to introduce the spectral view of signals, noise and signal recovery methods. The spectral view, in contrast to the previously adopted time-domain view, is to resolve all signals, whether wanted or unwanted, into sinusoidal spectral components. The main reason for this is that it is much easier to calculate the response of a circuit, or of any other form of processing, to a sine wave than to almost any other waveform.

So far the approach of the book has been unusual in using time-domain rather than frequency-domain methods. However, we have now reached the point where $1/f$ noise has to be treated, and this can only really be dealt with using the spectral methods. We start by introducing the general spectral approach, showing how both the required signal and random noise can be resolved into spectral components. The elementary processes of low-pass filtering and visual averaging, so far only covered by the time-domain method, are covered again from the spectral viewpoint; this serves to underline how the two approaches are just different ways of looking at the same basic effect. Then the methods developed are applied to $1/f$ noise in the following two chapters.

4.1 GENERAL METHOD

The essence of the spectral view is to resolve all signals, whether wanted or unwanted, into sine wave components. This is for two reasons. First, when a sine wave is applied to the input of a linear circuit (a circuit containing resistors, capacitors, inductors and linear amplifiers), the output is also a sine wave, differing only in phase and amplitude from that at the input. This contrasts with almost any other input waveform, for which the wave *shape* at the output will differ from that at the input. Second, and largely for the above reason, it is easier to calculate the response of a linear circuit to a sine wave than to any other waveform.

Signal spectrum
Figure 4.1 gives an example of the way in which a non-sinusoidal waveform can be resolved into sine wave components. The non-sinusoidal waveform is the square wave in figure 4.1(a). This, we shall see, can be resolved into the series of sine wave components

$$v = (4A/\pi) \sum_r r^{-1} \sin(2\pi f_0 rt) \tag{4.1}$$

where r is an odd integer and $f_0 = T_0^{-1}$. This spectrum is summarised in (h).

The first two components are shown in (b) and (c) and their sum in (e). Then the first three components are shown in (b), (c) and (d) and their sum in (f). Clearly, as the number of components increases the approximation to a square wave improves.

Low-pass filter
In the previous time-domain view, it has been shown that the essential function of a low-pass filter is to form a running average of the filter input. Figure 4.2 shows the effect of the filter in forming the running average of a sine wave of period T which is small compared with the averaging time T_r of the filter. Clearly the signal is considerably attenuated. Equally clearly, if $T \gg T_r$, the sine wave is transmitted with very little attenuation. Thus a filter cut-off frequency $f_r \sim T_r^{-1}$ can be defined. Then input components of frequency $f \ll f_r$ are 'passed' to the output, while components of frequency $f \gg f_r$ are rejected— hence the term 'low-pass filter'.

A more detailed analysis, as in §6.1, gives for the simple CR filter considered previously

$$f_r = (2\pi T_r)^{-1} \tag{4.2}$$

where $T_r = CR$. Note that the estimate $f_r \sim T_r^{-1}$ is in error by nearly an order of magnitude, so equation (4.2) should be used in preference to $f_r = T_r^{-1}$ even in rough approximations.

Filtering of square wave
From the spectral viewpoint, the effect of the filter is to remove those components of the square wave that lie above the filter cut-off frequency f_r. For example, for the frequency response shown in figure 4.1(h), all but the fundamental, third and fifth harmonics are rejected. Thus the filter output becomes as in (f). Further, if f_r is progressively reduced from the present $6f_0$ to $4f_0$, $2f_0$ and below f_0 the waveforms become as in (e), (b) and finally zero, as all components are rejected.

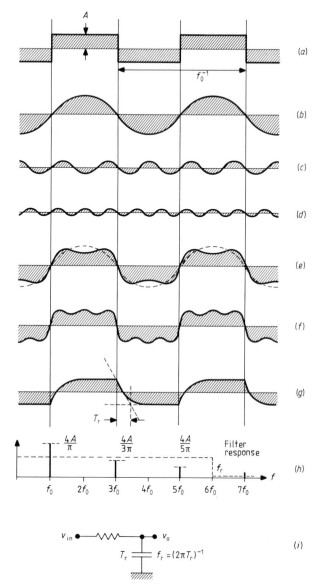

Figure 4.1 Fourier analysis of a square wave (a) into sine wave components: (b) fundamental of frequency f_0, (c) and (d) third and fifth harmonics, respectively, with frequencies $3f_0$ and $5f_0$, (e) fundamental + third harmonic, (f) with fifth harmonic added. (g) Square wave when filtered by low-pass filter (i). (h) Spectrum for square wave, showing frequency response of low-pass filter giving the output waveform in (f).

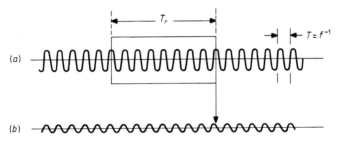

Figure 4.2 Attenuation of a sine wave input component (a) when subjected to a filter forming a running average over the period T_r (b).

The output waveform for a real filter is shown in figure 4.1(g). This is the result of applying a square wave to the simple single-section low-pass filter of (i). The waveform is a little different from the above predictions for two reasons. First, the cut-off of the real filter is not abrupt but gradual. Second, the real filter shifts the phase of each component to differing degrees. Of these, the phase shift is the more important effect.

For the present chapter such details need not concern us, and it will be sufficient to regard a low-pass filter as a device which transmits without attenuation in the pass band below the cut-off and totally rejects in the stop band above the cut-off. More detailed calculations are desirable for subsequent chapters and these are presented in chapter 6. The formal methods of Fourier analysis, the mathematical technique used to resolve a given signal into its spectral components, are given in the appendix.

Spectrum analyser

While Fourier analysis is the standard theoretical method for resolving a signal into its sine-wave components, the normal experimental method is to use a 'spectrum analyser' of the type shown in figure 4.3(a). This is a tunable narrow-band bandpass filter of frequency f_a followed by a detector which gives the output v_o which is proportional to the amplitude of the component at the filter output.

Thus, as f_a is varied at a constant rate over the spectral range of interest, the analyser output describes the spectrum of the signal at the analyser input. The diagram shows the square wave of figure 4.1 being analysed in this way. Clearly the bandwidth Δf_a of the filter determines the spectral resolution of the device.

4.2 SPECTRAL NOISE COMPONENT

Waveform

Figure 4.4 shows an experimentally observed spectral noise component. This is the waveform v_a observed at the output of the narrow-band bandpass filter

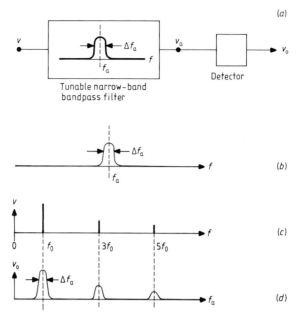

Figure 4.3 Use of a spectrum analyser (a) to resolve the square wave of figure 4.1 into its spectral components. (b) Frequency response of tunable filter, (c) spectrum of square wave signal, (d) spectral analyser output.

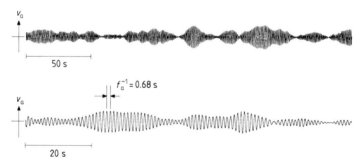

Figure 4.4 Experimentally observed spectral noise component of white noise as seen at the output of a narrow-band bandpass filter of central frequency $f_a = 1.5$ Hz and bandwidth $\Delta f_a = 0.05$ Hz.

in the spectrum analyser of figure 4.3 with white noise applied at the input. The essential features of this somewhat striking waveform can be explained by considering the effect of the filter on shot noise.

First, figure 4.5(a)–(c) shows the effect of one single pulse of the shot noise. The resulting output in (c) is basically the impulse response of the filter. The principal characteristics of this are well known. They are that the frequency of

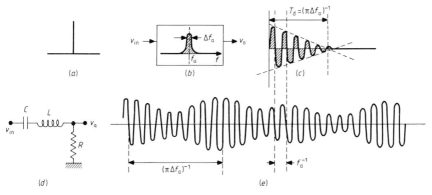

Figure 4.5 Diagram showing how the spectral noise component (*e*) at the output of the narrow-band filter in (*b*) is composed of the superposition of a random train of impulse response transients (*c*). (*a*) represents the input impulse; (*d*) shows the *LCR* resonator used as a filter in (*b*).

oscillation of the transient is equal to the filter frequency f_a and that the decay time T_d of the transient is equal to $(\pi \Delta f_a)^{-1}$, where Δf_a is the filter bandwidth.

For those unfamiliar with the results, the analogy of a mechanical resonator may be helpful. For example, a tuning fork has a resonant frequency and, if struck, will oscillate at that frequency with an amplitude that decays exponentially. The lower the damping the longer will be the decay time. In addition, if the fork is given a sinusoidal stimulus the response will be frequency selective. Moreover, the lower the damping the more selective will be the response.

If now the full sequence of pulses constituting the shot noise is applied to the filter input, each pulse will produce a corresponding output transient. Thus the output waveform will constitute the random superposition of many such transients. This produces a result of the type shown in figure 4.5(*e*), although with a greater amplitude than that shown. This is a sine wave of the frequency f_a of the filter, but fluctuating in amplitude with the fluctuation time equal to the decay time $T_d = (\pi \Delta f_a)^{-1}$ of the filter. The fluctuation time has this value because it is the time taken for each of the component transients in (*d*) to be replaced.

Phasor model
The model of figure 4.6 gives further insight. Here, for simplicity, each of the decaying transients of figure 4.5(*c*) is approximated by the flat-topped transients of figure 4.6(*a*). Figure 4.6(*b*) then shows the phasor that describes (*a*) on the real axis. Both phasor and pulsed sine wave have a 'lifetime' of $T_d = (\pi \Delta f_a)^{-1}$.

When a random series of such phasors is superimposed, the result is as shown in (*c*). Both resultant and component phasors rotate at the angular

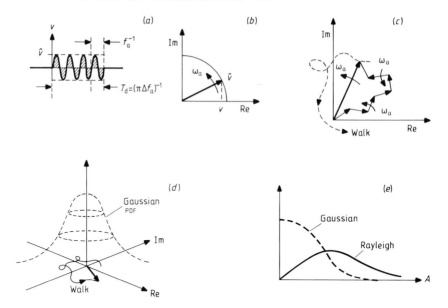

Figure 4.6 Diagram showing, from the phasor viewpoint, how the spectral noise component waveform in figure 4.5(d) arises from the superposition of many component transients. (a) Approximation to the transient of figure 4.5(c), (b) phasor describing this transient on the real axis, (c) superposition of many such phasors, (d) probability density function for random walk of composite phasor in (c) superimposed upon rotation at ω_a, (e) PDF for amplitude A of the phasor in (d) (Rayleigh distribution), compared with a gaussian.

frequency $\omega_a = 2\pi f_a$. Thus, if the resultant is viewed briefly from a frame of reference that rotates at the frequency ω_a, it will appear stationary. Soon, however, components will begin to be replaced and, as this happens, the tip of the phasor begins to 'walk'. After a time $(\pi \, \Delta f_a)^{-1}$, all of the components will have been replaced and then the phase angle of the new resultant will be entirely unrelated to the earlier one. Thus the time of walk of the resultant is equal to $(\pi \, \Delta f_a)^{-1}$.

We therefore have a resultant phasor that rotates at the mean frequency f_a with a superimposed slower fluctuation in amplitude, at the rate $\pi \, \Delta f_a$, and one can see how this will trace, on the real axis, the waveform of figure 4.5(e). A further point that emerges is that the frequency is not quite constant at the value f_a. In a period $(\pi \, \Delta f_a)^{-1}$, the phase walk is comparable with π and this corresponds to a frequency shift comparable with Δf_a. Thus the frequency fluctuates over the range Δf_a.

Statistical analysis allows the probability density function (PDF) for the noise component to be determined. This is shown in figure 4.6(d) and constitutes the familiar gaussian shape, rotated to form a 'bell'. The corresponding PDF for the amplitude A is shown in (e). This is the Rayleigh distribution and is simply the gaussian distribution multiplied by A.

The two PDFs differ because (d) represents the probability of the resultant phasor having a particular value on the complex plane. For any given amplitude A the range of such possible values is proportional to A, so that to obtain (e) the values in (d) must be multiplied by A.

4.3 PULSE SPECTRUM

In §4.6, where the spectral and time-domain views of signal recovery are finally compared, the signal used is a pulse, rather than the square wave considered above. The pulse spectrum is thus required. Once again the Fourier methods given in the appendix for a square wave are applicable. As for the square wave, a graphic presentation will be helpful and this is the object of the present section.

Figure 4.7(a) shows the flat-topped pulse and (b) its spectrum, as formally determined. In the graphic account we adopt the reverse procedure. Here the spectrum is assumed to have the idealised form of (d) and the corresponding waveform is shown to be that in (c).

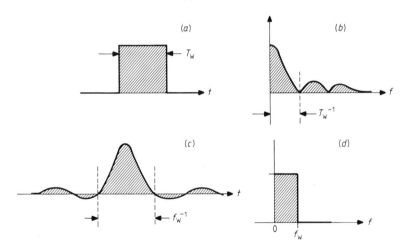

Figure 4.7 Waveforms and spectra for signal pulses. (b) and (d) are the spectra of the waveforms (a) and (c), respectively.

First, figure 4.8(a) shows the spectrum of figure 4.7(d) divided into a series of intervals of width δf, each containing one spectral component. Each component is assumed to be a cosine wave. Thus, in the phasor representation of (b), at $t=0$ the two component phasors will lie along the real axis. Then, as time proceeds, they will rotate in opposite directions at the characteristic frequency ω, as shown.

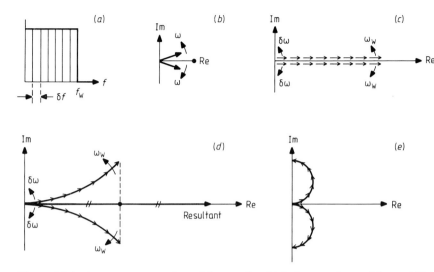

Figure 4.8 A phasor representation of the way in which the waveform of figure 4.7(c) can be recovered from a square wave spectrum (a). (b) Resolution of one of the spectral components of (a) into its phasor components, (c) and (d) the phasor sums at $t=0$ and $t>0$, respectively, (e) the first zero in the waveform, occurring when $t=(2f_{\mathrm{w}})^{-1}$.

Next, (c) shows each of the component phasors arranged in pairs at $t=0$. The angular velocity thus ranges from $\delta\omega$ for the lowest frequency component in (a) to $\omega_{\mathrm{w}}=2\pi f_{\mathrm{w}}$ for the highest frequency component. This will mean that at some little time later the diagram will become as in (d). Here, because of the linear increase in the angle described with frequency, the phasors will lie on a circle. This will mean that the resultant on the real axis will become zero at the time when the highest frequency phasors have rotated by π, as shown in (e), i.e. when $t=(2f_{\mathrm{w}})^{-1}$. This confirms the separation of the two zeros shown in figure 4.7(c) and also the general waveform. Clearly, as t advances, the resultant of the phasors, which is shown in figure 4.8(d), will describe the waveform of figure 4.7(c), as the value first decays to the first zero and then emerges to have a magnitude comparable with the diameter of a circle of ever decreasing value.

4.4 DEPENDENCE OF WHITE NOISE AMPLITUDE UPON BANDWIDTH

The particular property of white noise that gives it its name is that the amplitude of each of the spectral components is the same. The low-pass filter then reduces the noise by removing those components above the cut-off frequency f_{r} of the filter. The object of this section is to show that the amplitude $\tilde{v}_{\mathrm{n}}$ of the noise is proportional to $f_{\mathrm{r}}^{1/2}$. It turns out that the result is applicable to any kind of filter. Then $\tilde{v}_{\mathrm{n}} \propto B^{1/2}$, where B is the overall filter bandwidth.

Suppose, first, that the entire white noise spectrum is resolved into a series of spectral components of the type shown in figure 4.4, perhaps by using a series of narrow-band bandpass filters which equally divide the frequency range into increments of width Δf. In our analysis we shall assume that these are constant amplitude sine waves. This is permissible because, however long the measurement, Δf can be made sufficiently small for the amplitude fluctuation time $(\pi \Delta f)^{-1}$ of a component to be large compared with that of the measurement.

Let v_n be the sum of the noise components at the low-pass filter output. Then

$$v_n = \sum_r \hat{v}_r \cos \omega_r t \tag{4.3}$$

where v_r is the amplitude of the rth component. Then we may write

$$\overline{v_n^2} = \overline{\sum_r \sum_s \hat{v}_r \hat{v}_s \cos \omega_r t \cdot \cos \omega_s t} \tag{4.4}$$

$$= \sum_r \sum_s \tfrac{1}{2}\hat{v}_r \hat{v}_s \overline{[\cos(\omega_r + \omega_s)t + \cos(\omega_r - \omega_s)t]}. \tag{4.5}$$

Here, all averages are zero except when $\omega_r = \omega_s = 0$, giving

$$\overline{v_n^2} = \sum_r \hat{v}_r^2/2. \tag{4.6}$$

But $\hat{v}_r^2/2 = \overline{v_r^2}$ for a sine wave, so

$$\overline{v_n^2} = \sum_r \overline{v_r^2}. \tag{4.7}$$

So, for the n spectral noise components of equal amplitude

$$\overline{v_n^2} = n\overline{v_{\Delta f}^2} \tag{4.8}$$

where $\overline{v_{\Delta f}^2}$ is the value for any one component.

For an abrupt cut-off at f_r, $n = f_r/\Delta f$, giving $\overline{v_n^2} = (\overline{v_{\Delta f}^2}/\Delta f)f_r$. But both $\overline{v_{\Delta f}^2}$ and Δf are independent of f_r, so $\tilde{v}_n \propto f_r^{1/2}$. For the filter, $f_r = (2\pi T_r)^{-1}$ so $\tilde{v}_n \propto T_r^{-1/2}$. This confirms the dependence of $\tilde{v}_n$ upon T_r given by the time-domain analysis and indicates the consistency of the two views.

The above frequency-domain argument remains true whether the components come from a single continuous region of the spectrum or from different separated regions. It does not require specifically a low-pass response. Thus the result may be generalised to the expression

$$\tilde{v}_n \propto B^{1/2} \tag{4.9}$$

where B is the overall bandwidth of the filter. This ability to divide the spectral range is broadly equivalent to the time-domain result that the noise error is proportional to $T_{av}^{-1/2}$, where T_{av} is the averaging time, regardless of whether the averaging period is continuous, or separated as in the case of multiple time averaging.

4.5 NOISE WAVEFORM

Figure 4.9 provides an interesting comparison of the waveforms for a single spectral component and for low-pass filtered white noise. For the single component in (c) the characteristic frequency f_a is that of the bandpass filter in (a) accepting the component. The fluctuation time for the amplitude is equal to $(\pi \Delta f_a)^{-1}$ where Δf_a is the filter bandwidth. Thus, if Δf_a is increased as shown, f_a in (c) remains constant, but the amplitude fluctuation time $(\pi \Delta f_a)^{-1}$ is reduced. Finally, as Δf_a becomes $2f_a$, as in (b), the amplitude fluctuation time becomes comparable with the characteristic frequency and the transition from the relatively ordered waveform of (c) to the random one of (d) becomes complete.

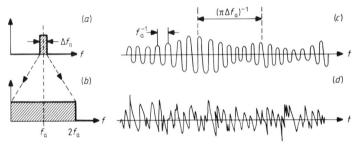

Figure 4.9 Change in the waveform of filtered noise from the single spectral component in (c), for the narrow-band bandpass filter of (a), to the random fluctuation of (d), as the bandpass is widened to that of the low-pass filter in (b).

Here the admission of additional spectral components as the bandwidth is increased will cause the output amplitude to increase. Thus, in reality, the amplitude of (d) will be a good deal greater than that of (c).

4.6 SPECTRAL AND TIME-DOMAIN VIEWS OF SIGNAL PULSE RECOVERY

We are now in a position to make a full comparison of the time- and frequency-domain views of the way in which a signal is recovered from white noise. The example used will be the recovery of the signal pulse shown in figure 4.10, and we shall consider the effects of low-pass filtering and visual averaging.

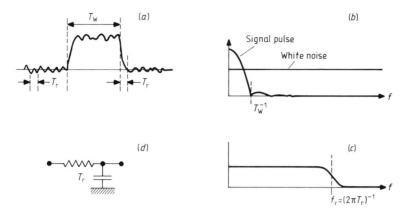

Figure 4.10 Waveforms and spectra for the low-pass filtering of a noisy signal pulse. (*a*) Waveform, (*b*) spectra, (*c*) frequency response of the low-pass filter (*d*).

Considering low-pass filtering first, the spectrum for the signal pulse is shown in (*b*), following figure 4.7(*b*). The white noise spectrum is also shown and the frequency-domain view of the way in which the filter reduces the noise is to say that the filter removes those components of the white noise spectrum that lie above the filter cut-off frequency $f_r = (2\pi T_r)^{-1}$ shown in the filter frequency response of (*c*).

The corresponding time-domain view is that the noise is reduced by the formation of an average over the period T_r of the filter. We can now compare the estimates of the limit of this reduction given by the two views. From the time-domain view, the filter averaging time T_r must not approach the width T_W of the pulse, otherwise the pulse will be distorted even more than is shown. From the frequency-domain viewpoint, the filter cut-off f_r must remain well above the upper limit T_W^{-1} of the signal spectrum. Since $f_r = (2\pi T_r)^{-1}$ for the simple low-pass filter shown in (*d*), the two estimates are in broad agreement. The factor of 2π by which the values differ arises from the somewhat arbitrary placings of cut-off points for averaging functions and frequency responses that are not abrupt. The agreement is good enough to be sure that we are discussing one effect from two different viewpoints and not two different effects.

It remains to consider the effect of visual averaging. This effect allows the averaging time T_r of the filter to be made less than the width T_W of the signal pulse because the eye is able to extend the averaging period up to the full pulse width. The waveform in (*a*) is drawn assuming this to be the case. Thus the noise fluctuation time, which is equal to T_r, is shown a good deal less than the pulse width T_W.

The spectral view of the visual averaging is that the eye averages out any spectral components that are of period identifiably less than the pulse width T_W. This is equivalent to imposing another low-pass filter, this time of cut-off frequency comparable with T_W^{-1}. But, from (*c*), this is the upper limit of the signal spectrum. Thus the CR filter rejects components down to f_r in (c) but the visual averaging reduces the limit to reject all noise components beyond the signal spectrum. It is thus again clear that the function of the CR filter is secondary and the precise setting of f_r unimportant.

4.7 SHOT NOISE AMPLITUDE

In these last two sections the classic expressions for the amplitude of filtered shot and thermal noise are derived. An expression for the amplitude $\tilde{v}_n$ of low-pass filtered shot noise has already been obtained in §1.2. Thus the present argument will be the frequency-domain counterpart of the earlier one and should confirm the earlier value.

The first step is to resolve the component shot noise pulse into its spectral components. Figure 4.11(a) shows one of these pulses. The method of Fourier analysis given in the appendix is applicable to periodic signals only, so, for the present purpose, it will be convenient to assume that the pulse repeats with the period T_0 as shown. Later on, T_0 can be allowed to become infinite so as to represent the single pulse.

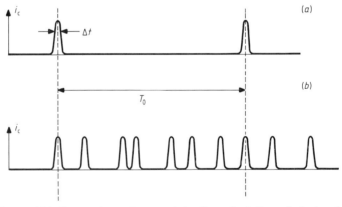

Figure 4.11 Shot noise pulses used in the calculation of shot noise amplitude: (a) single repeated pulse, (b) repeated random sequence.

For the periodic pulse train of (a) the spectrum will be a Fourier series of discrete spectral components separated by the interval T_0^{-1} on the frequency scale. Using the exponential form for the series,

$$i_c = \sum_{r=-\infty}^{+\infty} A_r \exp(2j\pi f_r t) \qquad (4.10)$$

where r is an integer, A_r is the value at $t=0$ of the rth Fourier component and $f_r = r f_0$, with $f_0 = T_0^{-1}$.

Then, taking the Fourier transform to evaluate the A_r,

$$A_r = T_0^{-1} \int_{-T_0/2}^{+T_0/2} i_c \exp(-2\mathrm{j}\pi f_r t)\,\mathrm{d}t. \tag{4.11}$$

i_c is only significant for $t \sim \Delta t/2$ to $-\Delta t/2$, and for this range the exponent is small compared with 2π, provided that $f_r \ll \Delta t^{-1}$. Then the exponential approaches unity and

$$A_r \simeq T_0^{-1} \int_{-T_0/2}^{T_0/2} i_c\,\mathrm{d}t \tag{4.12}$$

$$= q_e/T_0. \tag{4.13}$$

Thus, for frequencies small compared with Δt^{-1}, the spectral distribution is found to be uniform, as was demonstrated by the pictorial method of figure 4.8. The spectral density, moreover, is seen to be dependent only on the area of the shot noise pulse and not at all upon the detailed shape.

For frequencies higher than Δt^{-1} the exponential in equation (4.11) begins to oscillate over the pulse width Δt, and so the magnitude of A_r is reduced.

The next step is to include the effect of all of the other pulses that occur over the period T_0 shown in figure 4.11(b). Let p be the number of pulses over the period. Each pulse will have a corresponding spectrum, so that where there was previously only one component, now there are p. Because the pulses are timed randomly, the phasing of the components at any one frequency will also be random. Where p such components are added with random phasing, it can be shown that the probable amplitude of the resulting component is equal to $p^{1/2}$ times that for one of the components summed. Thus if A_{rp} represents the expected value of the sum,

$$A_{rp} = p^{1/2} A_r. \tag{4.14}$$

Each sine wave of frequency f_r is composed of two of the above exponential components, so that, if S_{rp} represents the expected magnitude of the sine wave of frequency f_r, then

$$S_{rp} = 2p^{1/2} A_r \tag{4.15}$$

and, from equation (4.13),

$$S_{rp} = 2p^{1/2} q_e/T_0. \tag{4.16}$$

The component is a sine wave, so that if $\overline{i_{rp}^2}$ represents the mean-square value of the component, then $\overline{i_{rp}^2} = S_{rp}^2/2$, giving

$$\overline{i_{rp}^2} = 2q_e^2 p/T_0^2. \tag{4.17}$$

Over a given frequency interval B the number of components to be summed is BT_0. Thus the mean-square noise current $\overline{i_{cn}^2}$ over B is given by

$$i_{cn}^2 = 2q_e^2 pB/T_0. \tag{4.18}$$

It remains to remove the periodicity of the pulse train in figure 4.10(b) by allowing the period T_0 to become infinite. This causes p to become infinite also, but if $\bar{r}$ is the long-term mean pulse rate, this will be independent of T_0. Then as $T_0 \to \infty$, $p \to \bar{r}T_0$ giving

$$i_{cn}^2 = 2q_e^2 \bar{r}B. \tag{4.19}$$

But $\bar{i}_c = q_e \bar{r}$, so

$$i_{cn}^2 = 2q_e \bar{i}_c B. \tag{4.20}$$

This is the classic expression for the amplitude of filtered shot noise. The corresponding expression $\sigma_n = q(\bar{r}/T_r)^{1/2}$ of equation (1.9), derived using the time-domain argument, gives the expected error σ_n due to shot noise averaged over the response time T_r of the filter and for the component pulse area q. Referring this expression to the current i_c makes $q = q_e$, $\bar{i}_c = \bar{r}q_e$ and $\tilde{i}_{cn} = \sigma_n$. Thus equation (1.9) becomes

$$i_{cn}^2 = q_e \bar{i}_c T_r^{-1}. \tag{4.21}$$

For equations (4.20) and (4.21) to be consistent requires $B = (2T_r)^{-1}$ for the filter. This is not quite in agreement with the relation $B = (2\pi T_r)^{-1}$ for a simple CR filter. However, it must be remembered that the two analyses proceed on different idealisations of the filter. The time-domain view assumes a box approximation to the memory function. This amounts to an abrupt cut-off in the time domain, while the spectral analysis assumes an abrupt cut-off in the frequency response. These two properties cannot be obtained simultaneously.

4.8 THERMAL NOISE AMPLITUDE

The object of the present section is to state the well known expression for the amplitude of filtered thermal noise and to give a simple and approximate verification of the validity of the expression.

The expression in question is that, if v_n is the noise voltage across the ends of a resistor R at temperature T as seen through a filter of bandwidth Δf, then

$$v_n^2 = 4kTRB \tag{4.22}$$

where k is Boltzmann's constant and $k = 1.38 \times 10^{-23} \, \text{J K}^{-1}$.

Actually, the expression is an approximation, valid only for frequencies below a certain limit. The thermal noise in the resistor is the low-frequency manifestation of the 'black body' radiation from the resistor. The understanding of this phenomenon is central to modern physics and was the initial factor leading to the revolution whereby classical mechanics was expanded into quantum mechanics. Figure 4.12 shows the black body curve, experimentally determined and accounted for by the quantum theory. This

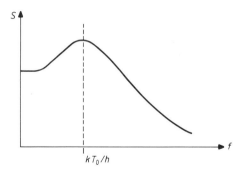

Figure 4.12 Spectral density S of black body radiation for frequencies in the region of kT_0/h.

shows that the frequency at which equation (4.22) becomes invalid is given by the relation $f = kT/h$, where h is Planck's constant ($h = 6.625 \times 10^{-34}$ J s), so that for normal temperatures, where $T = 300$ K, the critical frequency $kT/h = 6.3 \times 10^{12}$ Hz. Thus, equation (4.22) will be valid for all frequencies of current interest.

A general verification of equation (4.22) is beyond our scope, but one simple argument can be given.

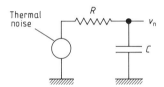

Figure 4.13 Circuit model used for the calculation of the amplitude of thermal noise for a resistor R.

In figure 4.13 the resistor is formed into its own low-pass filter by adding the capacitor C. Then the output from the equivalent noise generator as seen through the filter will be the capacitor voltage v_n shown. For any capacitor, the stored energy is equal to $\frac{1}{2}Cv_C^2$ where v_C is the capacitor voltage. The capacitor should be in thermal equilibrium with the resistor and therefore the mean energy stored should equal $kT/2$. Thus

$$C\overline{v_n^2} = kT. \tag{4.23}$$

But $f_c = (2\pi CR)^{-1}$, where f_c is the cut-off frequency of the filter, so

$$\overline{v_n^2} = 2\pi kTRf_c. \tag{4.24}$$

Since f_c in equation (4.24) corresponds to B in equation (4.22), the latter is confirmed, apart from the small factor $\pi/2$. This arises from the assumption in equation (4.22) that the cut-off is abrupt, while for the filter in figure 4.13 the cut-off is gradual.

5 $1/f$ noise

The main object of this chapter is to introduce the general properties of $1/f$ noise and to show, in particular, that, for a straightforward measurement, the $1/f$ noise error is independent of the time spent making the measurement. The example used is the previously encountered one of using the resistor bridge strain gauge of figure 1.1 to measure the stress–strain curve for a mechanical component. Here it is shown that the $1/f$ noise error is independent of the time T_{sc} taken to scan through the required range of applied stress.

At the end of the chapter a supporting experimental result is given. This mainly demonstrates the independence of $1/f$ noise error upon measurement time but also illustrates the effects of white noise and drift.

Chapters 6 and 7 then go on to show how the MTA and the PSD methods can be used to eliminate $1/f$ noise error and so realise the low white noise error obtainable by using a long period of measurement.

5.1 GENERAL PROPERTIES

Spectral distribution
The general appearance of $1/f$ noise was shown in figure 1.3(b), where it is compared with that of white noise. The $1/f$ noise has a greater tendency to 'walk away' from the mean and it is this property which prevents an increase in measurement time from giving a reduced noise error.

In fact, the names 'white' and '$1/f$' noise refer to the spectral distribution of the two types of noise. Figure 5.1 shows the use of a spectrum analyser to plot the noise spectrum for a simple wide-band feedback amplifier based on a commonly used operational amplifier with a JFET input stage. Here both white and $1/f$ noise are present and it is seen that the $1/f$ noise dominates at low

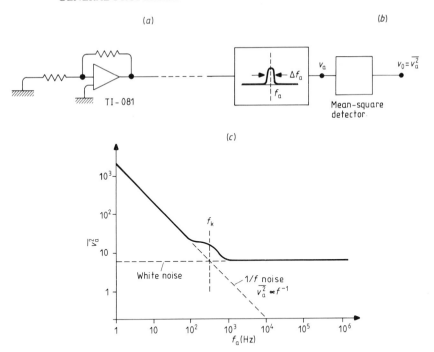

Figure 5.1 Spectral distribution of noise as recorded at the output of a typical wide-band feedback amplifier (a). (b) Spectrum analyser, (c) recorded noise spectrum.

frequencies and the white noise at high frequencies. The noise 'power' output $\overline{v_a^2}$ from the filter is independent of filter frequency f_a for the white noise but proportional to f_a^{-1} for the $1/f$ noise, hence the names. It is the preponderance of lower frequency components in $1/f$ noise that causes the walk away from the mean in the waveform of figure 1.3(b).

The transition frequency f_k in the noise spectrum of figure 5.1(c) is known as the '$1/f$ noise corner frequency' and will be of considerable importance subsequently. The small amount of additional noise to be seen in the region of f_k is a minor effect sometimes observed and need not concern us here.

Origins
In most cases $1/f$ noise can be said to originate from the imperfections in the manufacturing processes of an electronic component. Thus, in this sense, the $1/f$ noise can be regarded as 'extra', rather than 'basic', as for the white noise. Also, it may be expected that, as manufacturing techniques improve, the relative contribution of $1/f$ noise will be reduced.

In semiconductor device manufacture, the surface of the material is subject to a greater degree to imperfections in the crystalline structure, contamination, etc than the interior of the semiconductor slice. Thus any device that depends heavily upon surface conduction will be subject to a relatively large amount of 1/f noise. The notable example is the MOS transistor. For this device, conduction is largely over the surface, in contrast to the JFET and the bipolar transistor where the conduction is mainly beneath the surface. Thus the MOSFET exhibits a large amount of 1/f noise, and so is avoided in low-frequency designs. Figure 5.2 shows the noise spectra for two comparable operational amplifiers, one using MOSFET transistors and the other JFET and bipolar transistors. The difference in 1/f noise levels is apparent.

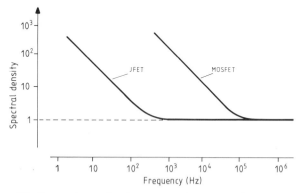

Figure 5.2 Comparison of noise spectra for JFET and MOSFET transistor operational amplifiers (spectral density shown relative to white noise level).

For much the same reason, small devices of any type are likely to be subject to a large amount of 1/f noise. This is because for a small device the ratio of surface area to volume of active area will be larger. Devices designed for high frequency operation are usually small, and particularly microwave transistors and point-contact rectifier diodes. Here, at the high frequencies for which the devices were designed, the increased 1/f noise will be of no consequence, being still well below the white noise. However, if the same devices are used for a low frequency application, the high 1/f noise level will be evident.

Some resistors exhibit 1/f noise, but only when a DC current is passed through them. This arises from fluctuations in the number of carriers present and therefore in the resistance of the device. The extent of the effect is material-dependent and tends to be more severe for cheap carbon composition resistors than when higher grade materials are used.

5.2 MEASUREMENT TIME INDEPENDENCE

The main object in this section is to show that 1/f noise error is independent of measurement time. The argument used involves some fairly gross approximations and will be superseded by a more general one in the next chapter. However, the present treatment allows the introduction of several concepts of more general importance.

The example used in the stress–strain curve measurement and it is shown that the various processes applied to the signal can together be approximated roughly by an ideal wide-band bandpass filter of upper and lower cut-off frequencies f_U and f_L. Thus the problem of estimating the noise error becomes one of calculating the effective output amplitude $\tilde{v}_n$ for such a filter.

Amplitude of bandpass filtered non-white noise

Suppose that the interval between the lower and upper limits f_L and f_U of the bandpass filter are divided into a series of smaller intervals of equal width Δf. Also let $v_{\Delta f}$ be the noise voltage for any one such interval. Then if v_n is the total noise voltage at the filter output,

$$\overline{v_n^2} = \sum_{f_L}^{f_U} \overline{v_{\Delta f}^2}. \qquad (5.1)$$

It is more convenient to have this expression in integral form. This requires the reduction of Δf towards zero. For a value of Δf sufficiently small, the noise over any one interval will be essentially white, although the noise over the full bandpass of the filter is non-white. For Δf below this limit, $\overline{v_{\Delta f}^2} \propto \Delta f$. Then $\overline{v_{\Delta f}^2}/\Delta f$ is independent of Δf and is termed the noise 'spectral density' G. Sometimes, for a noise voltage such as $v_{\Delta f}$, the value $\overline{v_{\Delta f}^2}$ is loosely termed the noise 'power'. Then G is the noise 'power' per unit bandwidth.

With G so defined, equation (5.1) becomes

$$\overline{v_n^2} = \sum_{f_L}^{f_U} G \, \Delta f \qquad (5.2)$$

and, since G remains constant as $\Delta f \to 0$,

$$\overline{v_n^2} = \int_{f_L}^{f_U} G \, df. \qquad (5.3)$$

For the spectrum-analysed $1/f$ noise in figure 5.1, $\overline{v_a^2} \propto f_a^{-1}$. But $\overline{v_a^2} = G \, \Delta f_a$, so $G \propto f_a^{-1}$, or, more generally, $G \propto f^{-1}$. This may be expressed as

$$G = A f^{-1} \qquad (5.4)$$

where A is a constant independent of f.

Then, from equation (5.3),

$$\overline{v_n^2} = A \int_{f_L}^{f_U} f^{-1} \, df = A \ln(f_U/f_L). \qquad (5.5)$$

For the stress–strain curve measurement, a low-pass filter is used, so that the cut-off frequency f_c of the filter constitutes f_U. However there appears to be no clear lower cut-off frequency, suggesting that $f_L = 0$. This creates difficulties because if in equation (5.5) $f_L = 0$ then $\overline{v_n^2}$ becomes infinite. The result is sometimes referred to as the 'low-frequency catastrophe' for 1/f noise, but is in reality illusory. This is because there is always some sort of low-frequency cut-off mechanism operating, and the problem is simply that of identifying it.

Baseline subtraction
Several such mechanisms exist and these will be reviewed more fully in the next chapter. For the present, we shall consider just one—the simple 'baseline subtraction' method of correcting offset which was shown in figure 1.11. For the moment, the method will be simplified to that of taking one point during the baseline period T_b and subtracting the value from any one of the required values of the strain-gauge output, as shown in figure 5.3(a). The diagram then shows the error caused by a spectral noise component (b) of frequency somewhat below that corresponding to the period in the region of T_{sc} which separates the two points. The precise value of the noise error is to some extent a matter of chance, depending on the time of occurrence of the scan relative to the noise component. However, even for the worst-case timing of (c), where the slope of the noise component is greatest, the noise error is less than the amplitude of the noise component. Moreover, as the frequency f_n of the noise component is reduced the ratio of the noise error to the component amplitude decreases further.

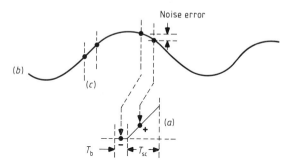

Figure 5.3 High-pass filtering effect of baseline subtraction. (a) Subtraction of baseline sample from signal sample, (b) low-frequency spectral noise component, (c) worst case timing.

In contrast, when $f_n > T_{sc}^{-1}$ the noise error for a given component amplitude will be comparable with the component amplitude, subject only to much the same type of uncertainty as for the lower frequency components. Thus the process of baseline subtraction has the essential features of a high-pass filter with a cut-off frequency comparable with T_{sc}^{-1}. The frequency response is analysed in detail in the next chapter and it is seen there that a good estimate for the cut-off frequency is the value $f_{sc} = (2\pi T_{sc})^{-1}$. This, then, is the value of f_L in equation (5.5).

Visual averaging

The other effect that should be taken into account in estimating the final noise error is that of visual averaging. If T_{res} is the required time resolution, then the corresponding upper cut-off frequency $f_U \sim T_{res}^{-1}$ and this would normally be lower than that of the filter. Again in the next chapter, it is shown that $(2\pi T_{res})^{-1}$ is a better estimate, so here $f_U = (2\pi T_{res})^{-1}$.

With $f_L = (2\pi T_{sc})^{-1}$ and $f_U = (2\pi T_{res})^{-1}$ in equation (5.5), and using σ_n in place of $\tilde{v}_n$, because this is now the expected error in a single measurement, the equation becomes

$$\sigma_n = [A \ln(T_{sc}/T_{res})]^{1/2}. \tag{5.6}$$

But T_{res}/T_{sc} is the fractional resolution χ, so

$$\sigma_n = [A \ln(\chi^{-1})]^{1/2}. \tag{5.7}$$

But both χ and A are independent of T_{sc}, and so it is finally shown that the final noise error σ_n is independent of scan time T_{sc}.

5.3 RESPONSE TO COMBINED 'NOISE'

It has now been established, at least for the example of the stress–strain curve measurement with baseline correction, that the $1/f$ noise error is independent of measuring time T_{sc}. We now consider the response to the 'combined noise' of white noise, $1/f$ noise and drift. It has been shown that for white noise the noise error $\sigma_n \propto T_{sc}^{-1/2}$, while the drift error increases with T_{sc}. The component noise errors are plotted in figure 5.4, as a function of T_{sc}, using logarithmic axes. In

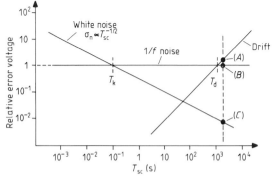

Figure 5.4 Errors due to white noise, $1/f$ noise and drift as a function of scan time T_{sc} in the stress–strain measurement.

practice the two transition periods T_k and T_d are separated widely as shown, so there is a wide range of values of T_{sc} for which the $1/f$ noise dominates and the total noise error is independent of T_{sc}.

The lower transition period T_k marks the change-over from white noise to $1/f$ noise dominance and clearly corresponds to the noise corner frequency f_k in the spectrum of figure 5.1(c). In fact, $T_k \simeq (2\pi f_k)^{-1}$.

The diagram shows how the low white noise error obtained in (C) by using a large value of T_{sc} is frustrated by the much larger drift and $1/f$ noise error in (A) and (B). It has been shown in earlier chapters that the drift error can be removed by either multiple time averaging or phase-sensitive detector methods. The question now is whether these methods are equally effective in removing the $1/f$ noise error. Fortunately the answer is 'yes' and one of the major objectives of the next two chapters will be to show how this is done.

5.4 EXPERIMENTAL RESULT

Figure 5.5 shows an experimental result that confirms the wide $1/f$ noise dominated region of the total noise error curve in figure 5.4. The result is actually not for the stress–strain curve of figure 5.4 but for the simpler step measurement of figure 1.8, but, this time, in the presence of combined noise rather than the totally white noise of figure 1.8.

The amplifier is that shown in figure 5.1(a) and which gives the combined noise spectrum of figure 5.1(c). The experimental procedure is much the same as for the steps observed in the presence of white noise alone. Thus the measurement period T_m is varied in decade steps, in this case over three decades. Also, as before, for each step, the recorder chart speed and the response time T_r of the low-pass filter are varied in proportion to T_m.

Over the range of values of T_m used there is almost no observable change in noise amplitude, thus indicating the $1/f$ noise dominance. The result of figure 5.5(e) for the largest value of T_m begins to show the effects of drift. This value of T_m therefore corresponds to the transition period T_d in figure 5.4. The result for the lowest value of T_m, shown in (b), still shows no increase due to white noise, therefore the transition period T_k will be less than this value, and in fact is almost a decade lower.

The results of figure 5.5 are in sharp contrast with those of figure 1.8 for white noise. For the white noise, the noise amplitude $\tilde{v}_n$ is seen to conform to the prediction $\tilde{v}_n \propto T_m^{-1/2}$, while for the $1/f$ noise, the independence of $\tilde{v}_n$ upon T_m is now also confirmed.

For the signal step, the high-pass filtering mechanism determining the lower cut-off frequency f_L is slightly different from that for the stress–strain curve measurement. For the step, baseline subtraction is not required, because any offset in the display, whether originating from true offset, drift or low-frequency $1/f$ noise components, can be ignored. The step measurement is

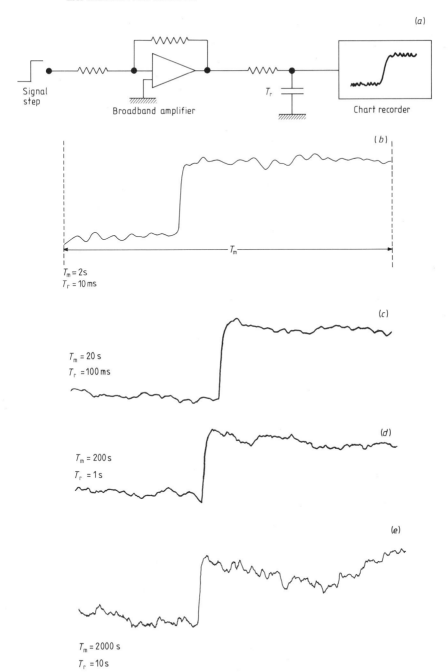

Figure 5.5 Measurement of low-level step in the presence of $1/f$ noise from the feedback amplifier of figure 5.1(a). (a) Experimental system, (b)–(e) results, showing that $1/f$ noise error is independent of measurement time T_{m}.

principally a matter of subtracting two values separated in time by $T_m/2$. Thus $f_L \simeq (\pi T_m)^{-1}$.

A noteworthy feature of the series of displayed noisy steps is that not only is the noise amplitude independent of T_m but so are the noise structure (i.e. the number of fluctuations over $T_m/2$), the fraction of the display width taken by the step response time T_r, and the final noise error σ_n. Indeed, apart from the drift in (e), the traces are almost entirely indistinguishable.

For the noise amplitude, f_U in equation (5.5) for $\overline{v_n^2}$ is equal to $(2\pi T_r)^{-1}$ and $f_L \simeq (\pi T_m)^{-1}$. But $T_r \propto T_m$, so $\overline{v_n^2}$ becomes independent of T_m.

The fluctuation time is T_r and so the number of fluctuations over $T_m/2$ will also be independent of T_m, since again T_r is varied in proportion to T_m. Finally, for the overall noise error σ_n, $f_L \simeq (\pi T_m)^{-1}$ as argued above and also $f_U \simeq (2\pi T_{res})^{-1}$. These values in equation (5.7) give $\sigma_n = [A \ln(2)]^{1/2}$, which also is independent of T_m.

6 Frequency response calculations

The main object of the last chapter was to show that, for a simple measurement, such as that of the stress–strain curve, the $1/f$ noise error is independent of measurement time T_{sc}. This involved approximating the various processes applied to the noisy signal by a wide-band bandpass filter of upper and lower cut-off frequencies f_U and f_L. Here f_L was associated with the simple baseline correction method, giving $f_L \sim T_{sc}^{-1}$, and f_U with visual averaging, backed by low-pass filtering, giving $f_U \sim T_{res}^{-1}$. Both of these relations are extremely approximate and it is the object of the first few sections of this chapter to calculate the various frequency responses with greater precision, thereby showing that the values $f_L = (2\pi T_{sc})^{-1}$ and $f_U = (2\pi T_{res})^{-1}$ finally used are better estimates.

These are not the only processes of interest to which there are corresponding frequency responses. To establish the normal method of calculating a frequency response, we start with the simple low-pass filter. This is followed by the somewhat similar process of forming a running average. Then comes the corresponding 'fixed' averaging process of visual averaging over the resolution time T_{res}. It is this last process which gives the upper cut-off $(2\pi T_{res})^{-1}$ above.

There are actually a good many different types of baseline correction that may be used to give a low-frequency cut-off and so to determine f_L. The more important of these are discussed in §§6.4 and 6.5. Finally, §6.6 discusses the case of a step measurement of the type shown in figure 5.5.

Section 6.7 presents an alternative argument showing that $1/f$ noise error is independent of measurement time. This is rather more general than the previous arguments and actually does not require detailed knowledge of the frequency response involved. The argument is based upon the concept of the 'spatial frequency' of a displayed noise component, i.e. the number of cycles per unit distance x across the display, rather than the normal 'temporal' frequency, which is the number of cycles per unit time.

The object of the last section is to calculate the frequency response of the multiple time averaging process. It is then possible to show how this can be used to avoid the $1/f$ noise in a real measurement, where both $1/f$ noise and white noise are present, and thus to realise the low white noise error that is obtained by using a long period of measurement.

6.1 LOW-PASS FILTER

The simplest frequency response to calculate is that of the single-section low-pass filter, which is shown in figure 6.1. This example will be used to introduce some of the methods to be used throughout the chapter.

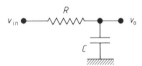

Figure 6.1 Simple single-section low-pass filter.

Essentially, the procedure used in order to calculate the frequency response of any network is to apply an input signal $v_{in} = \hat{v}_{in} \cos \omega t$ to the network and to calculate the values of $\hat{v}_o$ and ϕ for the resulting output signal $v_o = \hat{v}_o \cos(\omega t + \phi)$. The variation of $\hat{v}_o / \hat{v}_{in}$ with ω is then strictly termed the 'amplitude–frequency' response, and the variation of ϕ with ω the 'phase–frequency' response. Often, however, as in the present case, the variation of ϕ is of no interest and then it is normal to refer to the variation of $\hat{v}_o / \hat{v}_{in}$ with ω simply as the 'frequency response'.

To facilitate calculation, a slightly less direct approach is adopted. If the complex function $V = \hat{v} \exp[j(\omega t + \phi)]$ is plotted on the complex plane, as in figure 6.2, it may be shown to trace the circle indicated, rotating at the constant angular velocity ω. This vector-like function is sometimes called a 'phasor'. Then, from the diagram

$$V = \hat{v} \exp[j(\omega t + \phi)] = \hat{v} \cos(\omega t + \phi) + j\hat{v} \sin(\omega t + \phi). \qquad (6.1)$$

Thus the function $\hat{v} \cos(\omega t + \phi)$ can be thought of as the real component of the associated phasor V (in this chapter upper case notation will be used to denote a phasor).

Any such sinusoidal waveform has a corresponding phasor and thus the input component $v_{in} = \hat{v}_{in} \cos \omega t$ can be thought of as the real component of the input phasor $V_{in} = \hat{v}_{in} \exp(j\omega t)$. It is thus permissible to suppose that the actual input component v_{in} is replaced by the phasor V_{in}. In this situation it can then

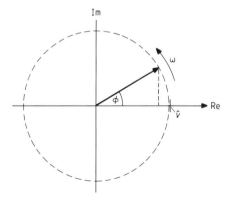

Figure 6.2 Phasor representation of $V = \hat{v} \exp[j(\omega t + \phi)]$, shown when $t = 0$.

be shown that all the network currents and voltages have the same basic exponential (phasor) form, including the output phasor V_o.

One of the main reasons for the indirect approach is that it makes calculation of the effect of the reactive components, such as the capacitor C, much easier. For the capacitor $i_C = C \, dv_C/dt$, which means that the shape of the current and voltage waveforms differ. If, however, the phasors I_C and V_C are used, then

$$I_C = C \, dV_C/dt. \tag{6.2}$$

So, we may write V_C and I_C as

$$V_C = \hat{v}_C \exp[j(\omega t + \phi_{vc})] \tag{6.3}$$

$$I_C = \hat{i}_C \exp[j(\omega t + \phi_{ic})] \tag{6.4}$$

which, with equation (6.2), gives

$$I_C = j\omega C V_C. \tag{6.5}$$

Thus the phasors V_C and I_C are related by the simple 'complex impedance' $X_C = (j\omega C)^{-1}$. The two waveforms are essentially of the same shape and the 90° phase difference is expressed by the operator 'j'.

Then, from the circuit diagram, $V_o/V_{in} = X_C/(R + X_C)$ and so

$$V_o/V_{in} = 1/(1 + j\omega/\omega_r) \tag{6.6}$$

where $\omega_r = (CR)^{-1}$.

Next, if H is written for the complex filter transfer function V_o/V_{in}, then $\hat{v}_o/\hat{v}_{in} = |H|$ and

$$|H| = 1/[1 + (\omega/\omega_r)^2]^{1/2}. \tag{6.7}$$

This is the filter frequency response which is plotted in figure 6.3. Strictly $|H|$

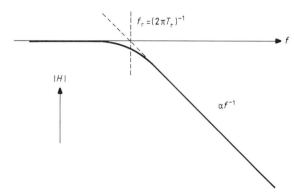

Figure 6.3 Frequency response curve for low-pass filter of figure 6.1 (log axes).

is the ratio of the magnitudes of the input and output phasors, but, from equation (6.1), this is also equal to the ratio of $\hat{v}_o/\hat{v}_{in}$.

Figure 6.3 is plotted using logarithmic axes as is usual. Here $|H|$ is plotted against f rather than ω. Thus the cut-off frequency $f_r = \omega_r/2\pi$. Also CR is equal to the response time T_r so $f_r = (2\pi T_r)^{-1}$ as shown.

6.2 RUNNING AVERAGE

An approximate time-domain view of low-pass filtering is to say that the filter forms the running average of the input, with the averaging period the time constant T_r of the filter. It is also possible to build a filter that actually does form the running average. For these reasons, it will be useful to calculate the frequency response of the true running-average-type filter and to compare it with that of the simple CR filter.

For the true running average filter of averaging period T_r

$$v_o(t) = T_r^{-1} \int_{t-T_r}^{t} v_{in}(t')\, dt' \tag{6.8}$$

as shown for the sine wave input component in figure 6.4(a).

Once again the phasors V_o and V_{in} are used in place of the real input and output components. Then, with $V_{in} = \hat{v}_{in}\exp(j\omega t)$,

$$V_o(t) = T_r^{-1} \int_{t-T_r}^{t} \hat{v}_{in}\exp(j\omega t')\, dt' \tag{6.9}$$

$$= V_{in}(t)\exp(-j\omega T_r/2)\,\frac{\sin(\omega T_r/2)}{\omega T_r/2}. \tag{6.10}$$

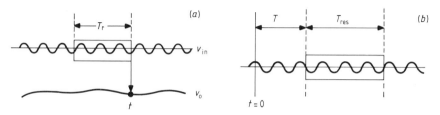

Figure 6.4 Comparison of (a) the formation of a running average over the period T_r and (b) the formation of a visual average over the resolution time T_{res}.

Then, since $|H| = |V_o|/|V_{in}|$,

$$|H| = \frac{\sin(\omega T_r/2)}{\omega T_r/2}. \tag{6.11}$$

This frequency response is plotted in figure 6.5, using both logarithmic and linear axes. The response for the simple CR filter of the last section is also shown for comparison. It is clear that, in this frequency-domain comparison, the running average is a reasonable approximation to the actual filter response.

6.3 VISUAL AVERAGING

The running average model of figure 6.4(a) can be adapted as in figure 6.4(b) to represent the process of taking the visual average of a single spectral noise component v_{in} over the required resolution time T_{res}. Here T represents the time interval between $t = 0$, at which the phase of the input component is zero, and the start of the averaging period. The value of v_o will vary sinusoidally with T in much the same way as the running average in (a) varies with t. However, for the fixed visual average, v_o has one value only and this will be somewhere between $\pm \hat{v}_o$, with the precise value depending upon where T places the averaging frame. Thus if the frequency response factor $|H| = \hat{v}_o/\hat{v}_{in} = |V_o|/|V_{in}|$ is calculated for the fixed average it must be remembered that the actual magnitude of v_o will be subject to a further reduction of, on average, about one half, relative to $|V_o|$.

With this reservation, $|H|$ can be calculated. Here only $|V_o|$ is of interest and T influences the phase rather than the magnitude of V_o. Thus T can be given any value that is convenient for the calculation. Here it is most convenient to make $T = 0$. Then

$$v_o = T_{res}^{-1} \int_0^{T_{res}} v_{in} \, dt \tag{6.12}$$

and with v_{in} replaced by the phasor $V_{in} = \hat{v}_{in} \exp(j\omega t)$,

$$V_o = \hat{v}_{in} \exp(j\omega T_{res}/2)\sin(\omega T_{res}/2)/(\omega T_{res}/2). \tag{6.13}$$

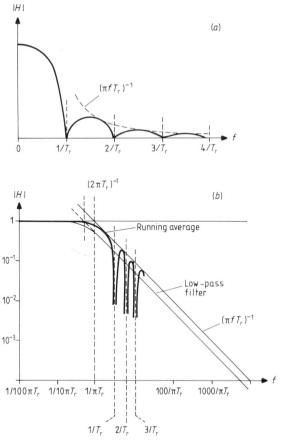

Figure 6.5 Frequency response for the running average of figure 6.4(a), compared with low-pass filter response. (a) Linear axes, (b) log axes.

But $|H| = |V_o|/|V_{in}|$ and $|V_{in}| = \hat{v}_{in}$, so

$$|H| = \sin(\omega T_{res}/2)/(\omega T_{res}/2). \qquad (6.14)$$

This is clearly identical in form to equation (6.11) for the running average, with T_r replaced by T_{res}. The frequency response of figure 6.5 will therefore also suffice for the visual average, with the cut-off frequency now $(2\pi T_{res})^{-1}$. This confirms the value of the upper cut-off frequency f_U used in the last chapter.

6.4 BASELINE OFFSET CORRECTION

There are several ways of correcting for baseline offset and in this section we calculate the corresponding frequency responses. All such responses are of the high-pass type, determining the lower cut-off frequency f_L in noise estimation, and in each case f_L will be calculated.

Single point correction
The first kind of correction to be analysed is that used in figure 5.3. This is the 'simple' baseline correction but with the further simplification that the baseline is sampled at one point only, rather than an average being taken over the period T_b. This preserves the essential high-pass filtering properties of baseline correction and constitutes the simplest possible demonstration of the point. It was this process that was used in order to predict the low-frequency cut-off f_L in the expression (5.5) for total $1/f$ noise error.

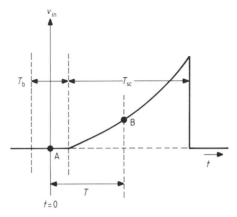

Figure 6.6 Baseline correction by subtracting the baseline value at point A from the signal + baseline value at point B.

Figure 6.6 shows again the essentials of the process. Here the baseline sample at A is subtracted from the signal sample at B. It is convenient here to place the time $t = 0$ at A. Then

$$v_o = v_{in}(T) - v_{in}(0). \tag{6.15}$$

With the single spectral input component represented by the phasor $V_{in} = \hat{v}_{in} \exp(j\omega t)$ as usual

$$V_o(t) = \hat{v}_{in}[\exp(j\omega T) - 1]. \tag{6.16}$$

But $|H| = |V_o|/|V_{in}|$ and $|V_{in}| = \hat{v}_{in}$, so

$$|H| = |1 + \exp(j\omega T)|. \tag{6.17}$$

This function is plotted in figure 6.7(a). The zeros at each multiple of $1/T$ arise because two samples taken from a sine wave spaced an integral number of sine wave periods apart will be the same. This is shown in figure 6.7(b).

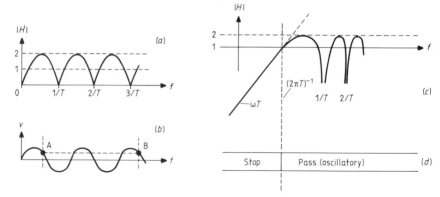

Figure 6.7(a) Frequency response for the baseline correction of figure 6.6. (b) Equal samples when the time between A and B is an integral multiple of the period of the input component. (c) Frequency response in (a) plotted using log axes. (d) Summarised frequency response.

That the frequency response is essentially that of a high-pass filter is clearer from figure 6.7(c), which is the same as (a) but plotted using logarithmic scales. For $\omega \ll T^{-1}$, $|H| \simeq \omega T$, which gives the asymptote shown. When, for the asymptote, $|H|=1$, $\omega=T^{-1}$ so $f=(2\pi T)^{-1}$, again as shown. This value appears to be a reasonable estimate of the cut-off frequency. Thus the stop band and pass band regions of the filter may be summarised as in (d). Here it is noted that the bandpass region is not smooth but oscillatory, varying as in (a) between zero and 2, rather than being stable at unity. From the overall noise transfer viewpoint this makes little difference.

From figure 6.6, it is clear that $T \sim T_{sc}$, although in fact T varies from $T_b/2$ to $T_{sc} + (T_b/2)$ depending upon which particular point of the stress–strain curve is being monitored. Thus $f_L \sim (2\pi T_{sc})^{-1}$ and this was the value used in equation (5.5) for the total $1/f$ noise error.

Simple baseline correction

We next calculate the frequency response for simple baseline correction of the more normal type, where the baseline sample is an average taken over the period T_b and the signal sample an average taken over the resolution time T_{res}. This is shown in figure 6.8(a) and we allow initially for the possibility of a delay T_d between the baseline sample and the stress scan. This is the situation when the baseline is evaluated, say, at the beginning of the day during which a series of measurements are to be made and the same value used throughout the day. It will be useful to see how much additional $1/f$ noise error results from this practice.

Often when the baseline is measured in this way the offset will be adjusted to bring the baseline error to zero, that is to some 'reference' datum line. This is

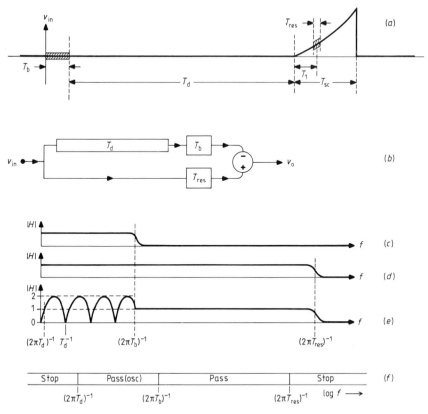

Figure 6.8(a) Simple baseline correction with additional delay T_d and averaging over T_b and T_{res}. (b) Block diagram representation of the process in (a) (assuming $T_d \gg T_b$ and T_{res}). (c)–(e) Frequency responses for upper, lower, and combined upper and lower arms of (b), respectively. (f) Summarised overall frequency response plotted to log frequency scale.

just a convenient way of subtracting the measured baseline and so the present analysis is equally applicable to this case.

Figure 6.8(b) summarises the process of figure 6.8(a), that of simple baseline subtraction, in a block diagram. The low-pass filters T_b and T_{res} represent the averaging over the respective periods. Then T_d represents a delay line which accounts for the delay between the two averaged samples.

Figure 6.8(c) then shows the frequency response for the upper arm of the pair. Here the delay line makes no difference to the amplitude so a simple low-pass characteristic results, with a cut-off frequency of $(2\pi T_b)^{-1}$.

Similarly the frequency response for the lower branch will be that of a low-pass filter of cut-off frequency $(2\pi T_{res})^{-1}$ as in (d).

Recall next that, in order to make the white noise error associated with the baseline sample small compared with that for the signal sample, it was required that $T_b \gg T_{res}$. This means that the cut-off frequency $(2\pi T_b)^{-1}$ of the baseline filter is well below that of $(2\pi T_{res})^{-1}$ for the signal averaging filter. Thus from the frequency domain viewpoint the way that the baseline noise error is made insignificant is by making the filter bandwidth small compared with that for the signal.

Next, figure 6.8(e) shows the combined response of the upper and lower branches. Above $(2\pi T_b)^{-1}$ the response is the same as for the lower branch alone. Below $(2\pi T_b)^{-1}$ in contrast, the two components interfere, because of the delay, much as in figure 6.7(a). Now $T = T_d$ so that the final summarised frequency response becomes as in (f).

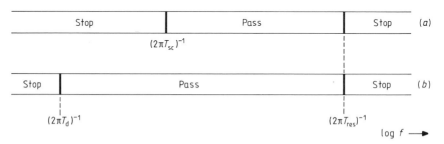

Figure 6.9 Summarised frequency responses when (a) T_d in figure 6.8(a) is reduced to zero and (b) $T_d \neq 0$ as in figure 6.8(a).

Figure 6.9 compares the summarised frequency responses when T_d is as in figure 6.8(a) and also when $T_d = 0$. When $T_d = 0$ the approximation $T_d \gg T_b$ and T_{res} no longer holds and the delay between the signal and baseline samples is effectively $T_b/2 + T_1$. This is in the region of T_{sc} as in the last section, so giving the lower cut-off of $(2\pi T_{sc})^{-1}$ shown in figure 6.9(a). Clearly, the effect of the delay is to extend the pass band in the lower frequency direction, thus accepting more $1/f$ noise.

For the purposes of evaluating the increase, it is convenient to have the frequency responses plotted on log axes, as in figure 6.9. This is because, from equation (5.5), the final noise error is proportional to $\ln(f_U/f_L)$, i.e. to the length of the pass band as displayed on the log scale. This means, for example, that the value of T_d for which the $1/f$ noise is doubled is given by $T_d/T_{sc} = T_{sc}/T_{res}$. This is χ^{-1}, where χ is the fractional resolution. Then, if χ is small, T_d can be made considerably in excess of T_{sc} before the total $1/f$ noise error is significantly increased.

Subtraction of signal average
In some kinds of measurement the mean of a displayed waveform is of no interest. One example is the measurement of a signal step. Here the interest is

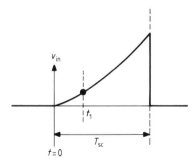

Figure 6.10 Subtraction of average over T_{sc} from each point on curve.

only in the difference between the values before and after the step; the mean is of no consequence.

Ignoring the mean is equivalent to subtracting it from each value. Were this to be done for the stress–strain curve in figure 6.10, the value of v_0 at the time t_1 shown would be given by

$$v_0(t_1) = v_{in}(t_1) - T_{sc}^{-1} \int_0^{T_{sc}} v_{in}\, dt. \tag{6.18}$$

Then for the input phasor $V_{in} = \hat{v}_{in} \exp(j\omega t)$

$$V_0 = \hat{v}_{in}\left(\exp(j\omega t_1) - \exp[(j\omega(T_{sc}/2)]\frac{\sin(\omega T_{sc}/2)}{(\omega T_{sc}/2)} \right) \tag{6.19}$$

and so

$$|H| = \left|1 - \exp[j\omega(T_{sc}/2 - t_1)]\frac{\sin(\omega T_{sc}/2)}{(\omega T_{sc}/2)}\right|. \tag{6.20}$$

This function depends upon t_1 but is the same for $t_1 = 0$ and $t_1 = T_{sc}$. The frequency response for these two cases is plotted in figure 6.11, together with that for the mid-way position $t_1 = T_{sc}/2$. This is a special case, being the condition for which not only baseline offset is cancelled but also constant-rate baseline drift. For all other values of t_1 only the offset is cancelled. The corresponding effect in the frequency domain is that when $t_1 = T_{sc}/2$ the low-frequency cut-off is of the second order, rather than of the first order, as is the case when $t_1 \neq T_{sc}/2$. A second-order cut-off is one where $|H| \propto f^2$, rather than $|H| \propto f$ as for the first-order cut-off.

The asymptotes and cut-off frequencies for the cases plotted are obtained in the usual way. For $t_1 = 0$ or T_{sc}, when $\omega \ll T_{sc}/2$, $|H| \simeq \omega T_{sc}/2$ and then $|H| = 1$ when $f = (\pi T_{sc})^{-1}$. When $t_1 = T_{sc}/2$ it can be shown that $|H| \simeq (\omega T_{sc}/2)^2/6$ and then $|H| = 1$ when $f = 6^{1/2}(\pi T_{sc})$.

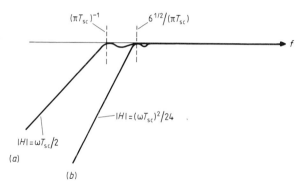

Figure 6.11 Frequency responses for figure 6.10 when (a) $t_1 = 0$ or T_{sc}, (b) $t_1 = T_{sc}/2$.

Switch-on effect

With all of the above possible mechanisms for establishing a low-frequency cut-off, it is still disconcerting that, when no method of correcting for offset is included, equation (5.5) suggests that, with the lower cut-off frequency $f_L = 0$, the total $1/f$ noise error $\sigma_n = \infty$. A realistic way of dealing with this problem is to conceptually divide the $1/f$ noise observed over a given period, such as the scan time T_{sc} in the stress–strain curve measurement, into two components, the mean and the variation relative to the mean. The variation component is then subject to a lower frequency cut-off and the $1/f$ model of the noise can successfully be used. The mean, in contrast, will be influenced by all manner of factors including the time for which the circuit has been switched on. It must thus be accepted that the $1/f$ model fails for this component.

6.5 SLOPING BASELINE CORRECTION

Figure 2.4 shows the method of sloping baseline correction. This was introduced primarily as a means of correcting for drift and actually entirely eliminates constant-rate drift. We now find the frequency response for this process and show that, in the low frequency bandstop region, the cut-off is a second-order one, i.e. $|H| \propto f^2$. This is true whatever the value of t_1, the time at which the signal sample is taken. This contrasts with the method of the last section, where the overall average is subtracted from each signal sample. There the cut-off is only of the second order type if $t_1 = T_{sc}/2$.

Figure 6.12 shows the time frame in detail. The baseline is reconstructed by taking averages over the two baseline periods T_b shown. These are termed $V_{b\alpha}$ and $V_{b\beta}$ and are also shown. Then the baseline is reconstructed using the relation

$$v_b(t_1) = (v_{b\beta} T_\alpha + v_{b\alpha} T_\beta)/(T_\alpha + T_\beta) \tag{6.21}$$

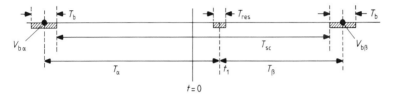

Figure 6.12 Time frame for sloping baseline correction of figure 2.4.

to give the value $v_b(t_1)$ of the baseline at any time t_1. This is then subtracted from the signal sample at t_1 to give the 'output' voltage for the process $v_o(t_1) = v_{in}(t_1) - v_b(t_1)$. If then v_{in} is replaced by the input phasor $V_{in} = \hat{v}_{in} \exp(j\omega t)$ the frequency response factor $|H|$ can be calculated as usual. The resulting expression is somewhat complicated but reduces to the approximation

$$|H| \simeq |\omega^2 (T_{sc}^2/8 - t_1^2/2)| \qquad (6.22)$$

when $\omega \ll T_{sc}^{-1}$.

This is second order regardless of t_1 but does depend somewhat on t_1. Taking the worst-case value of $t_1 = 0$, $|H| = 1$ when $f = 2^{1/2}/(\pi T_{sc})$.

This gives the lower frequency cut-off point in the summarised frequency response of figure 6.13. The upper cut-off and the transition frequency $(2\pi T_b)^{-1}$ arise much as in the simple baseline correction of figure 6.8. In the present case, the oscillatory nature of the lower part of the pass band results from the time difference between the signal sample period and the baseline averages.

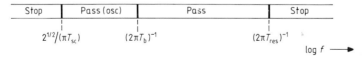

Figure 6.13 Summarised frequency response for sloping baseline correction.

6.6 STEP MEASUREMENT

The next frequency response to be analysed is that of the signal step shown in figure 6.14. This is of particular interest because the two experimental results so far presented have been of this type. Figure 1.8 showed such a series of measurements for a step in the presence of white noise and figure 5.5 the corresponding series for $1/f$ noise.

Here the process used in order to determine the step height is to subtract the average over the first half of the period of length T_m from the average over the second half. Thus

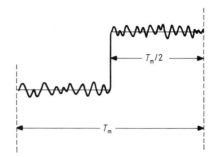

Figure 6.14 Noisy step.

$$v_o = 2T_m^{-1} \int_0^{T_m/2} v_{in}\, dt - 2T_m^{-1} \int_{-T_m/2}^0 v_{in}\, dt. \tag{6.23}$$

Then, with v_{in} replaced by the phasor $V_{in} = \hat{v}_{in}\exp(j\omega t)$ as usual

$$V_o = \frac{2\hat{v}_m}{j\omega T_m}\left[\exp(j\omega T_m/2) - 2 + \exp(-j\omega T_m/2)\right] \tag{6.24}$$

from which

$$|H| = \frac{\omega T_m}{2}\left(\frac{\sin(\omega T_m/4)}{\omega T_m/4}\right)^2. \tag{6.25}$$

This is plotted in figure 6.15, with the asymptotes and cut-off frequencies determined as usual. The important feature that emerges from this case is that the lower and upper cut-off frequencies are nearly equal. This makes the approximation of equation (5.5), the expression used to predict the $1/f$ noise error, suspect. Here it was assumed that the cut-off mechanisms were such as to give an abrupt cut-off at the limits f_L and f_U. For the present case this ignores

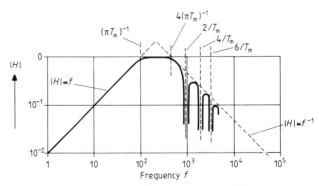

Figure 6.15 Frequency response for step in figure 6.14 when the averages over the two periods of length $T_m/2$ are subtracted.

most of the actual noise, which is in the 'wings' of the spectrum beyond the nominal cut-off points. A detailed calculation based on the true frequency response curve of equation (6.25) does actually confirm that the total $1/f$ noise error for this case is independent of T_m. However, it is somewhat tedious to have to establish this point for every case individually. It appears that there is some underlying principle at work which can establish that $1/f$ noise error is independent of measurement time regardless of the particular type of measurement. The next section constitutes an attempt to present such a generalised argument.

6.7 SPATIAL FREQUENCY COMPONENTS OF $1/f$ NOISE

We now present the more general argument, which is based on the concept of the spatial frequency components of $1/f$ noise, that the $1/f$ noise error is independent of measurement time. The example used is the measurement of a noisy signal step of the type shown in figure 6.16(a). This is typical of the series of experimentally recorded steps shown in figure 5.5. There the measurement time T_m was varied in decade steps over three orders of magnitude and the $1/f$ noise error found to be unaltered.

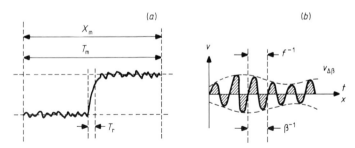

Figure 6.16 Measurement of noisy signal step. (a) Signal, (b) waveform of spectral noise component.

In this measurement the chart speed c and the response time T_r of the low-pass filter were varied in proportion to T_m. This makes the horizontal distance X_m shown on the display medium the same for each measurement. The distance occupied by the rise time T_r also becomes the same.

Figure 6.16(b) now shows a single noise component. This has the normal 'temporal' frequency f and also the 'spatial' frequency β. Here β is the number of cycles per unit display distance x, as opposed to the number of cycles per unit time, which is f. It is now possible to show that, when T_m and c are varied in the above manner, the magnitude of the noise at a given value of β and over a given small bandwidth, $\Delta\beta$, is unaltered.

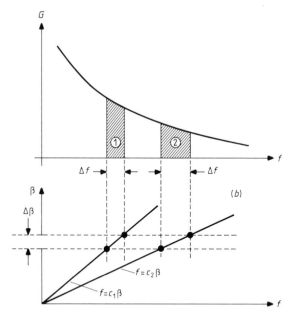

Figure 6.17 Diagram showing that the intensity of $1/f$ noise in a given interval of width $\Delta\beta$ and spatial frequency β is independent of chart speed c and measurement time T_m for the noisy signal step measurement of figure 6.16. (a) Spectral density for $1/f$ noise, (b) $\beta\text{–}f$ relation for chart speeds c_1 and c_2.

Figure 6.17 gives an idea of how this arises. Here (a) is the noise spectrum, plotted against f as usual. Now $f = c\beta$ and (b) shows this relation for two different values of c. It is seen that, for the given value of β, the noise is drawn from two different regions of the noise spectrum and also that the change in spectral density G tends to be offset by the change in bandwidth Δf. In fact the two effects exactly cancel, as will now be shown.

Since $f = c\beta$ then $\Delta f = c\,\Delta\beta$. Also, if Δv_n is the noise in the given interval at β, then $\overline{v_n^2} = G\,\Delta f$ and for $1/f$ noise $G = a/f$. Thus $\overline{v_n^2} = (A/c\beta)(c\,\Delta\beta)$. The c terms therefore cancel, giving

$$\overline{\Delta v_n^2} = (A/\beta)\,\Delta\beta \tag{6.26}$$

and the point is made.

It follows that, when c is altered as above, the noise component at each value of β is replaced by another, taken from a different part of the noise spectrum, but of the same amplitude. This makes the displayed structure of the noise, as well as the noise amplitude, the same for each measurement. These properties are shown clearly in the experimental results of figure 5.5, the only exception being (e), where some drift is evident. Clearly, with the variation of noise density with β independent of the chart speed c and the appearance of the signal, apart from noise, the same, the effect of any visually applied processing, such as forming averages over each period of length $T_m/2$, will be the same for each measurement. Thus it is fully established that the noise error, regardless of the type of filtering employed, is independent of measurement time T_m. Thus also we are relieved of the tedious task of calculating the exact frequency response of the filtering function used, in order to make our point.

6.8 FREQUENCY RESPONSE OF MULTIPLE TIME AVERAGING

In this section the frequency response of the multiple time averaging process is calculated and from this it is shown how the process can be used to eliminate $1/f$ noise error. Figure 6.18 shows the process applied to the stress–strain curve measurement. Here the applied stress is scanned repeatedly to produce the waveform shown in (a) at the strain gauge output. The outputs for the separate scans are then averaged to produce the result shown in (b), with the noise accordingly reduced. Here v_{in} in (a) is the 'input' to the multiple time averaging process and v_o in (b) the corresponding 'output'.

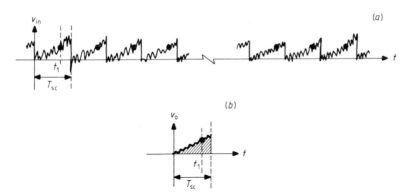

Figure 6.18 Multiple time averaging applied to the stress–strain curve measurement. (a) Strain gauge output, (b) average of n scans.

The value of v_o at the time t_1 in (b) is then given by

$$v_o(t_1) = n^{-1} \sum_{i=0}^{n-1} v_{in}(t_1 + iT_{sc}). \tag{6.27}$$

To calculate the response to one spectral input component, v_{in} is replaced by the usual phasor $V_{in} = \hat{v}_{in} \exp(j\omega t)$. This gives the output phasor

$$V_o = \hat{v}_{in} n^{-1} \sum_{i=0}^{n-1} \exp[j\omega(t_1 + iT_{sc})]. \tag{6.28}$$

It is convenient to extract $\exp[j\omega(t_1 - T_{sc})]$ from the sum to give

$$V_o = \hat{v}_{in}n^{-1}\exp[j\omega(t_1 - T_{sc})]\sum_{i=1}^{n}\exp(j\omega i T_{sc}). \tag{6.29}$$

But $|H| = |V_o|/|V_{in}|$ and $|V_{in}| = \hat{v}_{in}$ so

$$|H| = \left|\sum_{i=1}^{n}n^{-1}\exp(j\omega i T_{sc})\right|. \tag{6.30}$$

Figure 6.19 shows the components of the sum for a value of ω somewhat less than $(nT_{sc})^{-1}$. When $\omega = 0$, the sum lies along the real axis and $|H| = 1$. Then, as ω increases, the components curl around as shown, becoming a full circle, with

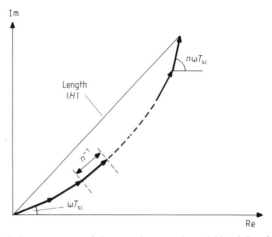

Figure 6.19 Components of the sum in equation (6.30) giving frequency response function $|H|$ for multiple time averaging.

$|H| = 0$, when $n\omega T_{sc} = 2\pi$, i.e. $f = (nT_{sc})^{-1}$. After a few minor oscillations $|H|$ becomes virtually zero until the condition $\omega T_{sc} = 2\pi$ is reached, i.e. $f = T_{sc}^{-1}$. Here all of the components have rotated by an integral multiple of 2π and so all are in phase again. This repeats for every multiple of T_{sc}^{-1} to give the frequency response of figure 6.20(a).

Since the repetition frequency for the signal shown in figure 6.18(a) is T_{sc}^{-1}, this has components at each multiple of T_{sc}^{-1} along the frequency scale, as shown in figure 6.20(b). These, not surprisingly, correspond to the acceptances in the frequency response of (a).

White noise reduction

We can now see how multiple time averaging reduces white noise. First figure 6.21(a) shows the effective frequency response for one scan of length T_{sc}. Here the cut-off frequency $(2\pi T_{res})^{-1}$ results from averaging, visually or otherwise,

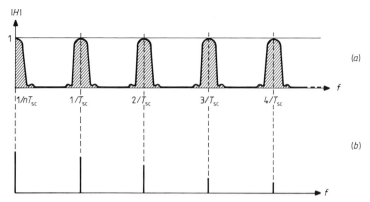

Figure 6.20(a) Frequency response $|H|$, (b) signal spectrum for multiple time averaging.

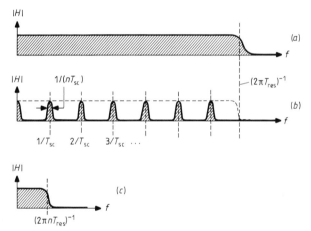

Figure 6.21 Comparison of frequency response curves for (a) a single short scan of length T_{sc}, requiring a resolution time of T_{res}, (b) multiple time averaging of n of the scans in (a), (c) a single long scan of length n times that in (a).

over the required resolution time T_{res}. There is, as yet, no $1/f$ noise so no lower frequency cut-off of the type provided by baseline correction is needed. Thus, the response is essentially a low-pass one as shown.

Next, (b) shows the result when n such scans are time averaged. Here the low-pass frequency response of (a) is combined with the multiple time averaging response in figure 6.20(a). The multiple time averaging thus reduces the noise by reducing the overall bandwidth. From the diagram, the bandwidth reduction factor is approximately n^{-1}, and from the noise reduction viewpoint it can be shown that the value is exact. But the white noise error is proportional to $B^{1/2}$, so, since the bandwidth B is reduced by n^{-1}, the noise error is reduced by $n^{-1/2}$. This is the same value as is given by time-domain analysis in §2.1.

$1/f$ noise reduction

Actually, for white noise alone there is no point in using multiple time averaging because, as will be confirmed shortly, the improvement in noise error that is obtained by time averaging n short scans is the same as that given by the simpler measure of increasing the scan time by the factor n and carrying out one scan only. Thus the real advantage of multiple time averaging lies when $1/f$ noise is present.

At first sight, the multiple time averaging frequency response of figure 6.21(b) does not look too promising as a means of avoiding $1/f$ noise. This is because there is an acceptance at zero frequency, which will accept not only $1/f$ noise but offset and drift. However it was established in chapter 2 that multiple time averaging is only effective in reducing offset and drift if it is combined with some sort of baseline correction mechanism. The same is true of $1/f$ noise and figure 6.22 shows simple baseline correction being used to avoid $1/f$ noise when using multiple time averaging. The cut-off frequency associated with baseline correction is $(2\pi T_{sc})^{-1}$ and it is seen that this suppresses the zero frequency acceptance and avoids the $1/f$ noise if the following two conditions are met.

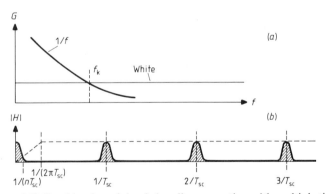

Figure 6.22 Combination of simple baseline correction with multiple time averaging in order to avoid $1/f$ noise. (a) Combined noise spectrum, (b) frequency response.

Firstly, in order to suppress the zero frequency acceptance, the baseline correction cut-off $(2\pi T_{sc})^{-1}$ must be well above the width $(nT_{sc})^{-1}$ of the zero frequency acceptance. This requires that $n \gg 2\pi$. Secondly, in order that the remaining acceptances are all in the region of the spectrum where the $1/f$ noise is insignificant, the scan frequency T_{sc}^{-1} must be above the noise corner frequency f_k.

Where the condition $n \gg 2\pi$ can only marginally be met, it will be of value to use sloping rather than simple baseline correction. This will give a second-order cut-off, rather than the present first-order cut-off.

With the $1/f$ noise thus removed, it appears that the low white noise error associated with a large period of measurement can now be realised. The time-domain diagram of figure 5.4 showed, for a single-scan measurement, how the various components of noise error vary with T_{sc}. The question then becomes whether the use of multiple time averaging can allow the low white noise error associated with the large T_{sc} and shown at point (C) to be realised. It has already been shown that multiple time averaging, combined with suitable baseline correction, removes offset and drift and now it has been established that $1/f$ noise can be removed also.

However, in order to be absolutely sure of realising the result, it must be established that, when the one long scan for (C) is replaced by n shorter scans of length reduced by n^{-1}, the white noise error remains the same. Actually this point has already been established by time-domain argument but it will now be of interest to obtain confirmation from the frequency-domain viewpoint.

Figure 6.21 (a) shows the frequency response for one short scan. Here T_{res}/T_{sc} is the fractional resolution χ which is independent of T_{sc}. Thus if T_{sc} is increased by the factor n so also is T_{res}. Thus the cut-off frequency $(2\pi T_{res})^{-1}$ is reduced by n^{-1} as in (c) and so, therefore, is the bandwidth. This is the same bandwidth reduction as is obtained in (b) by time averaging n of the shorter scans in (a) and so the noise reduction is the same. Thus the low noise error of point (C) in figure 5.4 is indeed realised.

Frequency-domain view of scan time increase
It is interesting to compare the time and frequency domain views of the result of increasing the scan time T_{sc} when white and $1/f$ noise are present. Figure 6.23(a) shows the combined noise spectrum and (b) shows the frequency response for one short scan. This is as for figure 6.21 (a) except that the $1/f$ noise makes baseline correction necessary. The associated cut-off frequency is $(2\pi T_{sc})^{-1}$ and, for the case shown, T_{sc} is assumed to be small enough for the entire pass band to be well above the noise corner f_k. Then, as T_{sc} is increased, both the lower and upper cut-off frequencies decrease in proportion to T_{sc}^{-1}. This gives a bandwidth reduction in proportion to T_{sc}^{-1} and therefore a noise error reduction in proportion to $T_{sc}^{-1/2}$. This corresponds to the left-hand side of the time-domain diagram of figure 5.4, where the white noise dominates.

As T_{sc} is increased further, past the point T_k on the time-domain diagram, the $1/f$ noise error becomes significant and no further improvement is obtained. This corresponds to the condition shown in (c) of the frequency response diagrams in figure 6.23. Here again the $1/f$ noise dominates and any reduction in bandwidth obtained by increasing T_{sc} is offset by the increase in noise spectral density G.

The above contrasts with the effect of leaving T_{sc} at the low value in figure

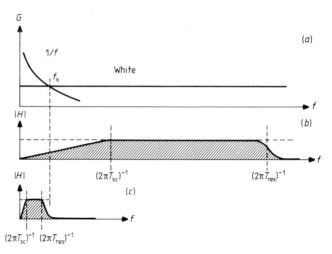

Figure 6.23 Reduction of scan time in the presence of white and $1/f$ noise. (a) Combined noise spectrum, (b) frequency response for single short scan using simple baseline correction, (c) frequency response for increased scan time.

6.23(b) and increasing the measurement time by multiple time averaging. Then the narrowing becomes, as in figure 6.22(b), in the region of the spectrum where $1/f$ noise is insignificant. Thus, in the time-domain diagram, the result is to hold to the $T_{sc}^{-1/2}$ line for the white noise error until eventually the point (C) is reached.

Practical limitations

One of the above requirements for multiple time averaging to avoid $1/f$ noise is somewhat unrealistic for the current example. This is that the scan frequency T_{sc}^{-1} should be above the noise corner f_k. Since $f_k \simeq 300$ Hz, the inertial limitations of any stress–strain testing machine would make the satisfaction of this condition impossible. Thus the present example is merely a means of illustrating the underlying principles.

For some applications, the noise corner is lower or the maximum repetition frequency is higher and then multiple time averaging can be used as suggested. For instances, such as the present example, where this is not so, it becomes necessary to use the alternative of the phase-sensitive detector method which is discussed in the next chapter. This, too, is an effective way of reducing $1/f$ noise, as well as drift and offset, and so allowing the low white noise error associated with a large measurement period to be realised.

7 Frequency-domain view of the phase-sensitive detector

The main objective of this chapter is to show that the phase-sensitive detector (PSD) method is capable of eliminating $1/f$ noise error. This requires the frequency-domain rather than the time-domain analysis of chapter 3, where it was shown that the PSD method effectively avoids errors due to offset and drift. With offset, drift and $1/f$ noise all avoided it is then possible to realise the low white noise error associated with a long period of measurement. In terms of the time-domain diagram of figure 5.4 this means that the low error of point (C) is obtained.

For the above conclusion to be valid it is necessary not only that the adaptation to the PSD method should eliminate offset, drift and $1/f$ noise but also that, for a given period of measurement, the adaptation should not increase the white noise error. This last point was argued somewhat loosely in chapter 3 using time-domain methods. With the present frequency-domain methods it is now possible to argue the point with greater rigour. This is the subject of §7.2.

There are actually several different types of PSD in common use and their suitability to different types of signal waveform is discussed in §§7.3–7.6.

Sometimes the PSD can be replaced by a simple non-phase-sensitive rectifier, notably when the signal-to-noise ratio at the detector input is large compared with unity. This is the subject of §7.7.

One of the valued functions of the PSD system is its ability to recover a signal that is totally submerged in noise at the detector input. This property is discussed in §7.8 and is one of the major differences between a phase-sensitive and a non-phase-sensitive detector system.

7.1 ANALOGUE MULTIPLIER PSD

Figure 7.1 shows again how the strain gauge is adapted to the PSD method. The simplest type of PSD for frequency-domain analysis is the analogue multiplier PSD of figure 7.1(b), rather than the reversing switch of chapter 3. Here the analogue multiplier is a device for which the output voltage v_p is proportional to the product of the two input voltages v_a and v_r. The waveforms of figure 7.2 for the analogue multiplier may be compared with those of figure 3.1 for the reversing switch in order to confirm that the analogue multiplier also performs the basic PSD function. Thus we analyse the analogue multiplier PSD first and leave the reversing switch to a later section.

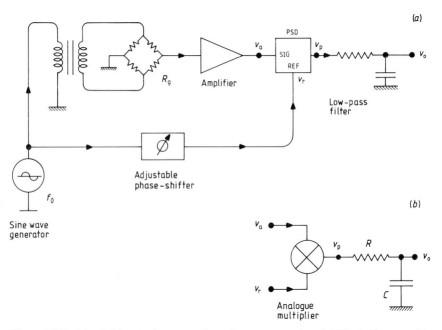

Figure 7.1 Resistor bridge strain gauge adapted to PSD operation. (a) Block diagram, (b) analogue multiplier PSD.

The important feature of the PSD from the point of view of $1/f$ noise avoidance is its spectral response. Figure 7.3(b) shows the signal spectrum and it is shown below that the input frequency response of the PSD matches the signal spectrum as shown in (c). Thus, provided the signal frequency f_0 is made somewhat greater than the noise corner f_k in the combined noise spectrum of (a), the $1/f$ noise is avoided.

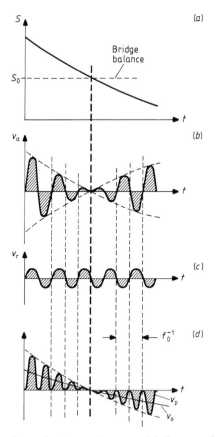

Figure 7.2 Waveforms for the analogue multiplier PSD of figure 7.1. (*a*) Strain, (*b*) signal input to PSD, (*c*) reference input to PSD, (*d*) PSD output waveforms.

Input acceptance pattern

We now derive the frequency response of figure 7.3(*c*) for the analogue multiplier PSD, referring to it, for reasons that will become clear shortly, as the 'input acceptance pattern' for the PSD.

The device is shown again in figure 7.4(α)(*a*). Then (α)(*b*) shows a single component at the signal input. This has the frequency f_s which will be allowed to assume any value and thus may represent any noise component or the required signal, for which $f_s = f_0$. It is the variation of the final output voltage v_o with f_s which constitutes the input acceptance pattern.

For the multiplier, with v_a, v_r and v_p as in the diagram and k_m the multiplier constant

$$v_p = k_m v_a v_r. \tag{7.1}$$

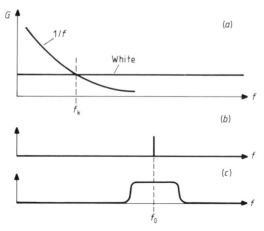

Figure 7.3 Use of PSD to avoid $1/f$ noise. (a) Combined noise spectrum, (b) signal spectrum, (c) required frequency response.

For the two input voltages:

$$v_a = \hat{v}_s \sin(\omega_s t) \tag{7.2}$$

$$v_r = \hat{v}_r \sin(\omega_0 t) \tag{7.3}$$

where $\omega_s = 2\pi f_s$ and $\omega_0 = 2\pi f_0$.
  Then

$$v_p = \tfrac{1}{2} k_m \hat{v}_s \hat{v}_r [-\cos(\omega_0 + \omega_s)t + \cos(\omega_0 - \omega_s)t]. \tag{7.4}$$

The spectra for v_s, v_r and v_p are shown in (b), (c) and (d), and it is seen that the effect of the two sinusoidal input components is to produce a further pair of sinusoidal components at the PSD output, with the output frequencies equal to the sum and difference of the two input frequencies.

Next consider the result of increasing f_s in (b) from zero to infinity. (e) shows the pass band of the low-pass filter following the PSD, with f_r the cut-off frequency of the filter. Then, as f_s passes through the range from $f_0 - f_r$ to $f_0 + f_r$, the component of frequency $f_0 - f_s$ passes from f_r, through zero, to $-f_r$. Here the minus sign has no significance, so it is for the full input range from $f_0 - f_r$ to $f_0 + f_r$ that the signal input component produces an output component accepted by the filter. For no value of f_s does the other output component in (d), that of frequency $f_0 + f_s$, fall within the pass band of the output filter. Thus the 'input acceptance pattern' of (f) represents the entire range of signal input frequencies for which a contribution is made to the final output v_o of the system, shown in (a). Thus the form of figure 7.3(c) is confirmed.

The reason why the term 'input acceptance' rather than 'frequency response' is preferred is now seen. The term 'frequency response' is used for such devices

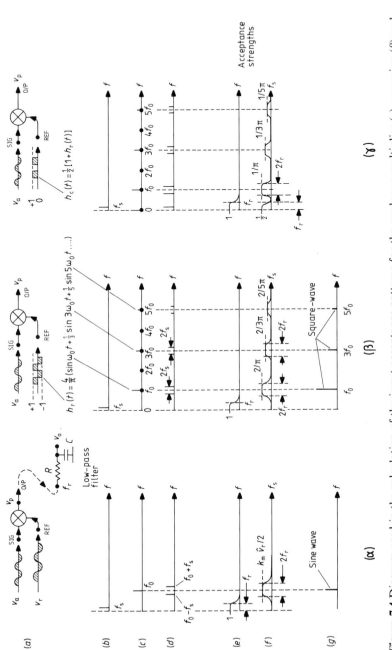

Figure 7.4 Diagram used in the calculation of the input acceptance patterns for the analogue multiplier (α), reversing (β) and chopping (γ) PSD circuits. (a) PSD, (b)–(d) spectra for v_a, v_r and v_p, (e) frequency response of low-pass filter, (f) input acceptances, (g) spectrum of required signal. See text for further details.

as filters which produce only output components of identical frequency to those at the input. The PSD, in contrast, involves the process of frequency conversion.

The value $k_m \hat{v}_r / 2$ shown in the input acceptance pattern of figure 7.4(α)(f) is known as the 'acceptance strength' and is defined as the ratio of the amplitude of a component of the final PSD output v_o, to the amplitude of the signal input component giving rise to the output component. The value here is given by equation (7.4).

7.2 WHITE NOISE ERROR

With the above results it is now possible to show more precisely than before that, when a simple system like the strain gauge bridge of figure 1.1 is adapted to the PSD method, the white noise error for a given period of measurement is in no way altered. This is done here for the adaptation of figures 7.1 and 7.2, where the bridge is driven by an AC supply and an analogue multiplier is used as the PSD. This is somewhat easier to analyse from this viewpoint than the adaptation of figure 3.1, where a square-wave generator and a reversing PSD were used. This was the adaptation used for the time-domain argument of the same point using figure 3.8. This is covered later in §7.3, once the acceptance pattern for the reversing PSD has been derived.

Figures 7.5(a) and (b) show the necessary features of the two systems for the comparison of white noise signal-to-noise ratios. Here (a) is the 'elementary' or unadapted strain gauge of figure 1.1. The bridge and amplifier are grouped together and it is assumed that there is a strain induced bridge imbalance producing the output signal E_{se} given by $E_{se} = E_g k_{ba}$ in which k_{ba} is a constant summarising the effect of the bridge and amplifier.

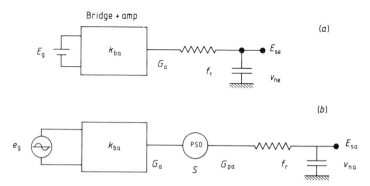

Figure 7.5 Summary diagrams used to show that the white noise signal-to-noise ratio is unaltered when the basic strain gauge in (a) is adapted to phase-sensitive detection as in (b).

G_a represents the spectral density of the white noise at the amplifier output and v_{ne} the resulting noise voltage at the final output. With f_r the filter cut-off, $\tilde{v}_{ne} = (G_a f_r)^{1/2}$. Thus the signal-to-noise ratio $(S/N)_e$ for the elementary system will be given by

$$(S/N)_e = E_g k_{ba}/(G_a f_r)^{1/2}. \tag{7.5}$$

Figure 7.5(b) then shows the system adapted to the PSD method as described. In order to make a fair comparison, the same amount of power should be dissipated in the bridge in both cases. This will require that $\hat{e}_g = 2^{1/2} E_g$, where e_g and E_g are the two generator voltages.

From the diagram, $E_{sa} = \hat{e}_g k_{ba} S$, where S is the acceptance strength of the PSD. Thus $E_{sa} = 2^{1/2} E_g k_{ba} S$. Here E_{sa} is the signal component at the output of the adapted system.

It remains to calculate $\tilde{v}_{na}$ where v_{na} is the noise voltage at the output of the adapted system in (b). If G_{pa} is the noise spectral density at the PSD output then $G_{pa} = 2S^2 G_a$. The factor 2 occurs because, from the acceptance pattern for the analogue multiplier PSD shown on figure 7.6(b), it is clear that for each small interval on the frequency response of the filter at the PSD output there are two corresponding intervals on the input acceptance.

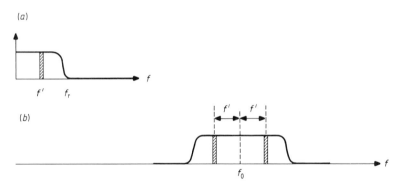

(a)

f' f_r

(b)

f' f'

f_0

Figure 7.6 Diagram showing that the noise component at any given frequency f' at the PSD output originates from two input components. (a) Output filter response, (b) input acceptance pattern.

Then, since $\tilde{v}_{na} = (G_{pa} f_r)^{1/2}$, it follows that $\tilde{v}_{na} = (2S^2 G_a f_r)^{1/2}$.

The signal-to-noise ratio $(S/N)_a$ for the adapted system is then given by $E_{sa}/\tilde{v}_{na}$, which becomes

$$(S/N)_a = E_g k_{ba}/(G_a f_r)^{1/2} \tag{7.6}$$

and this is the same as $(S/N)_e$ for the elementary system in equation (7.5). Thus it is shown that adaptation of the elementary strain gauge system to the PSD method makes no difference to the white noise signal-to-noise ratio.

The main objective of the present chapter, and indeed of the first part of the book, has now been achieved, in that it has now been shown that the PSD method is capable of avoiding offset, drift and $1/f$ noise errors without degrading the white noise error. Also, in the last chapter it was shown that multiple time averaging could be used to the same end. Thus it is established that by either of these methods, in principle, the low white noise error exemplified by point (C) in the time-domain diagram of figure 5.4 for a long period of measurement can be obtained.

As stated above, the remaining sections in this chapter cover the other types of PSD sometimes used, the use of non-phase-sensitive detectors and the use of the PSD to recover a signal buried in noise.

7.3 REVERSING PSD

We now consider the spectral view of the reversing PSD, for which the waveforms are shown in figure 3.1. First the input acceptance pattern is calculated. This is shown in figure 7.4(β)(f) for comparison with that of the analogue multiplier in (α). Next it is shown that for this system, too, there need be no reduction in the white noise signal-to-noise ratio and also that $1/f$ noise can be avoided. It is in this section that the concept of matched filtering mentioned on the overview begins to emerge.

Input acceptance pattern
The derivation of the input acceptance pattern for the reversing PSD follows much the same line as for the analogue multiplier PSD and figures 7.4(α) and (β) compare the two developments. It is possible to represent the reversing function by the arrangement of figure 7.4(β)(a). This is an analogue multiplier of unity gain constant, with the reference input the 'reversing function' $h_r(t)$ also shown. It should be understood that the arrangement is merely a convenient model for analysis; it is not proposed for actual use.

From the diagram,

$$v_p = h_r v_a. \tag{7.7}$$

Then h_r can be resolved into the Fourier series also shown, i.e.

$$h_r = 4\pi^{-1} \sum_{n=1}^{\infty} n^{-1} \sin(2\pi n f_0 t) \tag{7.8}$$

where n is odd, positive and integer.

The signal voltage v_a in figure 7.4(β)(a) remains the input sine wave given by equation (7.2) which, with equations (7.7) and (7.8), gives

$$v_p = 2\hat{v}_s \pi^{-1} \sum_{n=1}^{\infty} n^{-1}\{-\cos[2\pi(nf_0 + f_s)t] + \cos[2\pi(nf_0 - f_s)t]\}. \tag{7.9}$$

Here, as before, when f_s is in the range from $f_0 - f_r$ to $f_0 + f_r$ the component of v_p of frequency $f_0 - f_s$ is accepted by the low-pass filter. Now, however, when f_s is in the range from $3f_0 - f_r$ to $3f_0 + f_r$, the output component of frequency $3f_0 - f_s$ is accepted by the filter—and so on, for each component of frequency $nf_0 - f_s$ at the PSD output. Thus the acceptance pattern and strengths become as in figure 7.4(β)(f).

White noise error
We next confirm that the reversing PSD, when used with a square-wave signal, gives the same output signal-to-noise ratio as that for the 'elementary' system. Here figures 3.8(a) and (b) compare the waveforms for the two systems. There it is clear that the signal voltage at the output of each system is the same. Thus it will be sufficient to show that the output noise levels are the same. But the same output filters are used, so it will also be sufficient to show that the noise spectral densities at the filter inputs are the same. Here, if G_a is the noise density at the signal amplifier output and G_{pr} the density at the PSD output, then $G_{pr} = 2\sum_{n=1}^{\infty} S_n^2 G_a$ where S_n are the acceptance strengths for the various acceptances and n is odd. The factor of 2 is, as before, because for any one acceptance, there are two input components contributing to each output component (as was demonstrated using figure 7.6).

With the acceptance strengths as given in figure 7.4(β)(f)

$$G_{pr} = 2G_a(2/\pi)^2[1 + 3^{-2} + 5^{-2} + \ldots]. \tag{7.10}$$

But the series $1 + 3^{-2} + 5^{-2} + \ldots$ is equal to $\pi^2/8$, so that it is established that $G_{pr} = G_a$ and thus that the signal-to-noise ratio for the reversing PSD adaptation is the same, for white noise, as for the original unadapted system.

The extra acceptances for the reversing PSD are all above that at f_0 so that again, provided f_0 is made larger than the noise corner frequency, the $1/f$ noise is avoided. Also, since it is now clear that the same white noise error is given as for the unadapted system, the reversing PSD is equally effective as a means of realising the low white noise error associated with a large measurement time.

Waveform and spectrum matching
There is one reservation that must be made concerning the above comment. While both the analogue multiplier and the reversing PSD circuits have been shown to be capable of giving the same white noise error as for the unadapted system of figure 1.1, it should be noted that for the analogue multiplier a sine wave signal was assumed, while for the reversing PSD a square wave was assumed.

The corresponding signal spectra are shown in figure 7.4(g) and it is seen that the signal spectra 'match' the acceptance patterns in (f). This is essentially because the signal waveforms match the reference waveforms in (a). The reference waveforms then produce the reference spectra in (c) and it is these which determine the input acceptance patterns in (f).

It is shown later that if this 'matching' is violated, say by using the reversing PSD with a sine wave signal or by using the analogue multiplier with a square wave signal, then the white noise error is somewhat increased relative to that for the 'elementary' system prior to PSD adaptation. It thus appears that such matching is necessary for minimum white noise error. It is also clear that matching of signal and reference waveform is the time-domain view of the criterion, while matching of signal spectrum and input acceptance is the frequency-domain view.

7.4 CHOPPING PSD

Another type of PSD that is sometimes used is the simple chopping switch shown in figure 7.7(f). The waveforms in (a)–(e) make it clear that the basic function is fulfilled by the device. It will be seen shortly however that the chopper has some disadvantages relative to the other two types of PSD discussed so far. Thus, some reason will be required for using the device at all.

The reason for using the chopping PSD is its simplicity. This means it can be made to operate at higher signal frequencies than either of the other two types. The analogue multiplier is difficult to realise for operating frequencies above a few MHz and the reversing PSD allows operating frequencies of up to only about 50 MHz. The chopper, in contrast, can be made to operate for values of f_0 in the GHz range.

Input acceptance pattern
In order to compare the chopping PSD with the other two types, the input acceptance pattern must be determined. The development is illustrated in figure 7.4(γ) and follows the same lines as for the other two devices. In (a) the chopping function h_c replaces the reversing function h_r for the reversing PSD. Clearly

$$h_c = (1 + h_r)/2. \tag{7.11}$$

Also $v_p = h_c v_a$ so with h_r as given in equation (7.8),

$$h_c = \tfrac{1}{2} + 2\pi^{-1} \sum_{n=1}^{\infty} n^{-1} \sin(2\pi n f_0 t) \tag{7.12}$$

where n is odd, positive and integer.
Then, as before,

$$v_p = \frac{\hat{v}_s}{2} \sin(2\pi f_s t) + \hat{v}_s \pi^{-1} \sum_{n=1}^{\infty} \{-\cos[2\pi(nf_0 + f_s)t] + \cos[2\pi(nf_0 - f_s)t]\} \tag{7.13}$$

and the corresponding input acceptance pattern and strengths become as shown in figure 7.4(γ)(f).

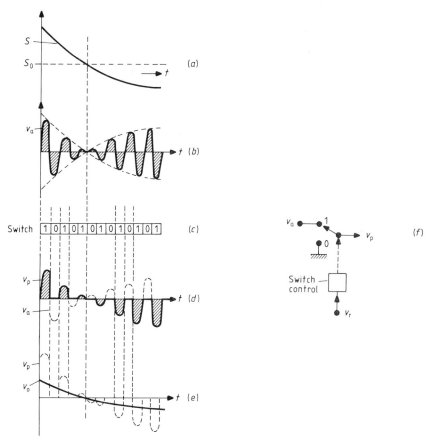

Figure 7.7 Waveforms for chopping PSD. (*a*) Strain, (*b*) signal input to PSD, (*c*) switch state, (*d*) PSD output, (*e*) low-pass filter output, (*f*) chopper.

The main disadvantage of the chopping PSD is now clear. This is the zero frequency acceptance which will accept offset, drift and $1/f$ noise. The waveform of figure 7.8 confirms this point. Here the voltage v_a applied to the PSD input constitutes offset and constant-rate drift. The combined effect of the chopping switch and the final low-pass filter is to reduce this by one half. This contrasts with the equivalent waveforms for the reversing switch, which are shown in figure 3.7. Here the offset and drift are rejected altogether.

A further disadvantage of the chopping switch is that it 'wastes' the signal for one half of the time. This causes an increase in noise error relative to the output signal by $2^{1/2}$. Both of these disadvantages can be overcome by preceding the PSD by a high-pass filter with a cut-off of approximately $f_0/2$. This removes the zero frequency acceptance and so avoids the offset, drift and $1/f$ noise. There is no signal component corresponding to this acceptance, whether the signal is a

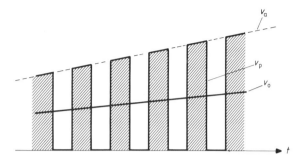

Figure 7.8 Response of chopping PSD to offset and drift at input.

sine or a square wave, so removal of the white noise associated with the acceptance will tend to compensate for the increased white noise error. In fact, the compensation is exact and so the white noise error becomes the same as for the reversing PSD.

To show that the reduction in noise error obtained by removing the zero frequency acceptance is by $2^{-1/2}$, exactly compensating for the $2^{1/2}$ increase imposed by replacing the reversing PSD by the chopper, note that, in the main, the acceptance strengths for the chopper are one half of those for the reversing PSD. For the reversing PSD the spectral density at the output and the input are the same. Therefore the spectral density at the chopper output is one quarter of that at the input, considering only the acceptances other than the zero-frequency acceptance.

For the zero-frequency acceptance, the acceptance strength is one half. Therefore the output spectral density corresponding to this acceptance is one quarter of that at the input. This is the same as that due to all of the other acceptances and so the effect of removing the zero frequency acceptance is to reduce the noise voltage by $2^{-1/2}$.

Pre-filtering

The above technique of adding a pre-filter at the PSD input in order to remove unwanted acceptances can sometimes be usefully extended. In particular, when the signal is a sine wave, calling for the analogue multiplier type PSD, it is possible to use either a chopper or a reversing switch and remove the unwanted acceptances by a simple bandpass filter tuned to the signal frequency f_0. This will be particularly useful when the operating frequency is such that an analogue multiplier cannot be realised but even at lower frequencies the high cost of an analogue multiplier may make the alternative arrangement attractive.

7.5 COMPROMISE ARRANGEMENTS

A good deal has been made about the desirability of matching the PSD response to the signal waveform, by using an anlogue multiplier type of PSD for a sine wave signal and a reversing switch for a square wave. Indeed, it has been shown that these two combinations give no loss in white noise signal-to-noise ratio relative to the 'elementary' strain gauge prior to adaptation to the PSD method. We have not, however, calculated the loss incurred by violating this matching, and in fact the loss is quite small. Thus in this section we calculate the loss for the two 'crossed' combinations of an analogue multiplier PSD with a square wave signal and a reversing PSD with a sine wave signal. Of the two, the most likely to be used is the analogue multiplier with the square wave signal. This is advantageous in the presence of relatively strong interference components that are of fixed frequency. Clearly, it is easier to avoid these by suitable adjustment of f_0 if the system has only one input acceptance rather than many.

Reversing PSD with sine wave signal

The loss in signal-to-noise ratio when a reversing PSD is used with a sine wave signal, rather than the 'matched' analogue multiplier, is now calculated. When the signal is a sine wave, having the signal spectrum shown in figure 7.4(α)(g), one way of obtaining the matching single-acceptance pattern of the analogue multiplier, shown in (α)(f), is to use a reversing PSD preceded by a bandpass filter designed to accept only the acceptance centred on f_0 in the acceptance pattern shown in (β)(f) for the reversing PSD.

If then the prefilter is removed, the reduction in signal-to-noise ratio will equal that resulting when the compromise arrangement of the unmodified reversing PSD is used with the sine wave signal.

The spectrum of the sine wave signal has no components within the higher order acceptances of the reversing PSD acceptance pattern. Thus removal of the pre-filter makes no difference to the output signal level, and the sole reason for the decrease in signal-to-noise ratio is the additional noise admitted by the higher acceptances.

Equation (7.10) gives the spectral density G_{pr} of the white noise at the output of the reversing PSD system with no prefilter. Here the terms in the series $1 + 3^{-2} + 5^{-2} + \ldots$ correspond to the various acceptances in the acceptance pattern of the PSD. Therefore, the increase in output noise power when the prefilter is removed will be by the factor $1 + 3^{-2} + 5^{-2} + \ldots = \pi^2/8$.

Removal of the filter makes no difference to the output signal level, so the reduction in output signal-to-noise ratio is $8^{1/2}/\pi \simeq 0.9$.

Analogue multiplier PSD with a square wave signal

When a PSD having the input acceptance pattern of the analogue multiplier is used with a square wave signal, there are two factors influencing the reduction in output signal-to-noise ratio. Firstly, the single acceptance pattern of the analogue multiplier rejects all of the components of the square wave signal spectrum apart from the fundamental of frequency f_0. Secondly, the noise over the higher order acceptances in the multiple acceptance pattern of the reversing PSD is rejected. To calculate the overall loss in signal-to-noise ratio, we assume that the PSD is initially of the reversing type, and calculate the reduction in both output signal and noise levels when the PSD is preceded by the bandpass filter necessary to convert the multiple acceptance pattern of the reversing PSD to the single acceptance pattern of the analogue multiplier.

To calculate the reduction in signal level, let E_{sr} and E_{srf} be the DC signal voltages at the output of the reversing PSD system without and with the prefilter. Also, let E_{sa} be the zero-to-peak value of the square wave signal at the output of the signal amplifier. To determine E_{srf} the square wave is resolved into its Fourier components. Then the prefilter accepts only the fundamental component, of frequency f_0. This, from §4.1, is equal to $(4/\pi)E_{sa} \sin \omega_0 t$.

The PSD then rectifies this sine wave, to give a signal that repeats with the period $(2f_0)^{-1}$. The following low-pass filter then forms the average of this rectified waveform, which constitutes the final signal output E_{srf}. Thus

$$E_{srf} = 2f_0 \int_0^{1/2f_0} (4/\pi)E_{sa} \sin \omega_0 t \; dt \qquad (7.14a)$$

$$= 8E_{sa}/\pi^2. \qquad (7.14b)$$

Then figure 4.7 shows that, without the prefilter, the reversing PSD and square wave signal give a final DC output voltage equal to the amplitude of the square wave signal. Thus, since in the present analysis E_{sa} is the signal amplitude and E_{sr} the output signal voltage, $E_{sr} = E_{sa}$.

Then, with equation (7.14), the ratio E_{srf}/E_{sr} of the signal voltages with and without the prefilter is given by

$$E_{srf}/E_{sr} = 8/\pi^2. \qquad (7.15)$$

It was shown above that, when the prefilter is removed, the increase in output noise power is by $\pi^2/8$. Thus, when the filter is replaced, the reduction in output noise amplitude will be by $8^{1/2}/\pi$.

Then, with equation (7.15), the reduction in signal-to-noise ratio when the prefilter is added is by $8^{1/2}/\pi \simeq 0.9$. Thus, for both of the compromise arrangements discussed, the loss in signal-to-noise ratio is the same and, as stated, quite small.

7.6 PSD IMPERFECTIONS

Looking at the input acceptance patterns in figure 7.4, it is clear that the patterns match the reference spectra in (c). In principle, the reference might have any waveform and this would, in general, give a reference spectrum with all multiples of f_0 present, including zero frequency. For a real PSD circuit of any of the types shown, the reference waveforms will not be exactly that required and the deviations will produce minor components at zero frequency and also at the even multiples of f_0, for which ideally there are no components. The result is that a real PSD will have minor or 'vestigial' acceptances at those multiples of f_0 for which the ideal PSD has none.

Zero frequency acceptance

The most serious of these vestigial acceptances is that at zero frequency. This is because such an acceptance will make the PSD susceptible to input offset drift and $1/f$ noise. One cause of such a zero frequency offset has already been mentioned. In §3.3 it was observed that, for the reversing PSD, if the reference switching duty-cycle is not exactly 1:1, or if the gains in the forward and reverse switch states are not quite equal, this will allow some transmission of drift and offset. It is precisely this kind of effect that causes the zero frequency acceptance. The way to avoid the acceptance is the same as for the chopping PSD, for which there is a major zero-frequency acceptance. The PSD is simply preceded by a high-pass filter with a cut-off frequency in the region of $f_0/2$.

In general, the even-order acceptances, or even the unwanted odd-order acceptances in what is supposed to be an analogue multiplier response, are less serious. The amount of extra white noise accepted will be negligible and the only possible difficulty will be strong interference components. Once again, a suitable prefilter will overcome this difficulty.

7.7 NON-PHASE-SENSITIVE DETECTORS

Figure 3.1(g) shows what happens when the reversing PSD in (a) is replaced by the non-phase-sensitive full-wave rectifier of figure 3.2. The distortion of the strain function only occurs if the bridge balance is traversed and this can be avoided by applying a sufficiently large initial imbalance. With this provision, the phase of the required signal does not reverse and the phase-sensitive detector is not required.

In this section we explore the possibility of replacing the PSD by the equivalent non-phase-sensitive circuit. First it will be useful to note what are the non-phase-sensitive equivalents of the three types of PSD discussed so far.

When the bridge balance is not traversed, the waveforms in figures 3.1(f) and (g) for the reversing PSD and the full-wave rectifier are identical. Thus the full-wave rectifier is the non-phase-sensitive counterpart of the reversing PSD.

Figure 7.9 shows the waveforms for the half-wave rectifier. Again, if the bridge balance is not traversed, the waveforms become exactly the same as for the chopping PSD in figure 7.7. Thus the half-wave rectifier is the non-phase-sensitive counterpart of the chopping PSD.

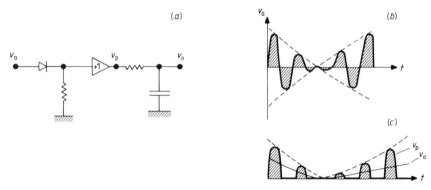

Figure 7.9 Waveforms when half-wave rectifier (*a*) replaces the chopping PSD in figure 7.7: (*b*) input, (*c*) output.

This leaves the analogue multiplier PSD. The counterpart for this is the squarer shown in figure 7.10. Here again, provided the bridge balance is not traversed, the waveforms for the squarer become very similar to those shown in figure 7.3 for the analogue multiplier. There is one important difference, however, and this is that v_o for the squarer is proportional to the square of the amplitude $\hat{v}_s$ of the sine wave signal input and not to $\hat{v}_s$ directly.

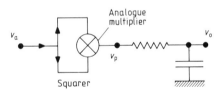

Figure 7.10 Mean square detector circuit $v_o = \overline{v_a^2}$.

In fact, for the squarer, $v_o = \overline{v_a^2}$ and the device is a mean square detector. There are some applications where this is desirable. For example, a noise spectrum analyser usually requires an output proportional to $\overline{v_n^2}$, since it is to this that the noise spectral density G is proportional.

Auto-referencing
In each of the non-phase-sensitive detector circuits the reference function of the phase-sensitive counterpart is provided by the signal. For example, for the analogue multiplier PSD the reference waveform is a sine wave. For the squarer, one of the multiplier inputs can be thought of as the signal input proper and the other the reference. Thus the reference is the sine wave signal.

Notice also that for the full-wave rectifier of figure 3.1(*g*) the diodes are switched by the signal input at exactly the same time as the reversing switch PSD in (*f*) changes state. Thus, once again the signal acts as a reference and the same sort of comparison can be made for the half-wave rectifier and the chopping PSD.

It is this 'auto-referencing' that gives the insight into the reason for one of the major limitations of the non-phase-sensitive detector. This is that the device will only operate successfully if the noise level at the signal detector input is small compared with that of the signal. For significant input noise levels, the reference becomes less than ideal. Then the switch timing for the full- and half-wave rectifier circuits becomes incorrect and also the reference waveform for the squarer becomes corrupted.

It will by now be clear that the 'matched' non-phase-sensitive detector for a sine wave signal is the squarer and for a square wave signal the full-wave rectifier. Provided the input signal-to-noise ratio is high, these will both give the same white noise output signal-to-noise ratios as for the phase-sensitive counterparts and therefore for the original system, prior to adaptation to the AC bridge drive. Only when the input signal-to-noise ratio becomes poor does the non-phase-sensitive detector give a poorer result than its phase-sensitive counterpart.

It will also be clear that all of the prefiltering strategies applicable to the phase-sensitive detector systems are applicable to the non-phase-sensitive devices. For example, the half-wave rectifier, like the chopping PSD, has a zero-frequency acceptance which accepts input offset, drift and $1/f$ noise. This must be removed by a suitable high-pass filter. Likewise, when the signal is a sine wave and the use of an analogue multiplier is not desirable, it is possible to use a full-wave rectifier preceded by a bandpass filter centred on the signal frequency f_0. An important point concerning this variant is that it gives an output proportional to the signal amplitude $\hat{v}_s$ and not to $\hat{v}_s^2$ as for the squarer. For most applications this will be preferred.

Signal-to-noise ratio loss
Figure 7.11 gives another view of the way in which the output signal-to-noise ratio for a non-phase-sensitive detector becomes less than that for the phase-sensitive counterpart when the input signal-to-noise ratio is not high. The example chosen is a square wave signal, for which the reversing PSD is replaced by the full-wave rectifier. (a) shows the rectifier and (b) the waveform for the rectifier output v_p when the peak value $\hat{E}_s$ of the square wave is large compared with the amplitude $\tilde{v}_n$ of the noise at the rectifier input. This is identical to the waveform for the reversing PSD shown in figure 3.8, and so it may be assumed that the same signal-to-noise ratio at the final output will be obtained as for the PSD.

The final output $v_o = \overline{v_p}$ and for $\hat{E}_s \gg \tilde{v}_n$, $\overline{v_p} = \hat{E}_s$. Thus $v_o = \hat{E}_s$, giving the variation of v_o with $\hat{E}_s$ shown in (e), for large values of $\hat{E}_s$.

It is when $\hat{E}_s$ is reduced to approach $\tilde{v}_n$ that the simple rectifier begins to fail. (c) shows the waveform for v_p in the extreme where $\hat{E}_s = 0$. Then $v_o = \overline{v_p}$ is comparable with $\tilde{v}_n$ and the full variation of v_o with $\hat{E}_s$ becomes as shown in (e).

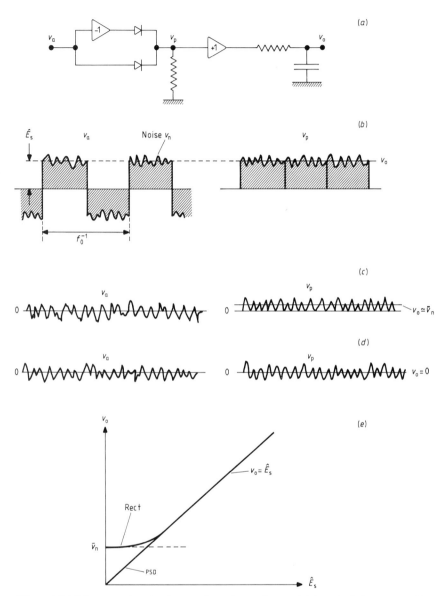

Figure 7.11 Diagrams showing how noise reduces the sensitivity of a full-wave rectifier to a small signal of amplitude less than that of the noise. (a) Full-wave rectifier, (b) waveforms when signal amplitude $\hat{E}_s \gg \tilde{v}_n$, (c) waveforms when $\hat{E}_s = 0$ for full-wave rectifier, (d) waveforms when $\hat{E}_s = 0$ for reversing PSD, (e) variation of detector output v_o with signal amplitude $\hat{E}_s$ for full-wave rectifier and reversing PSD.

This may be contrasted with the waveform of (d) for the output v_p of the reversing PSD when $\hat{E}_s = 0$. Here $v_o = 0$. Thus the v_o, $\hat{E}_s$ relation differs for the phase-sensitive and non-phase-sensitive devices as shown in (e).

It is when $\hat{E}_s \ll \tilde{v}_n$ that the full-wave rectifier is particularly at a disadvantage. The noise fluctuation components in v_o are comparable for both types of detector but the signal component for the rectifier is much less than that for the PSD. This is because of the much reduced slope of the v_o, $\hat{E}_s$ relation. Thus it is shown, albeit qualitatively, that for an input signal-to-noise ratio $\hat{E}_s/\tilde{v}_n < 1$, the non-phase-sensitive full-wave rectifier is considerably inferior to the phase-sensitive reversing switch. The same sort of argument can be advanced for the other types of non-phase-sensitive detector.

7.8 RECOVERY OF A SIGNAL SUBMERGED IN NOISE

One of the properties that distinguishes a phase-sensitive detector from its non-phase-sensitive counterpart is that the PSD is able to resolve a signal that is buried in broad-band noise of an amplitude many times that of the signal. Simply by reducing the bandwidth of the output filter, the bandwidth of the input acceptance is similarly narrowed. This rejects most of the input noise components and so the signal is recovered. This contrasts sharply with the corresponding non-phase-sensitive detector for which successful operation requires that the signal-to-noise ratio at the detector *input* be large compared with unity.

Limit to input signal-to-noise ratio

It seems that, in principle, there is no limit to how low the input S/N ratio can be for the input signal to be resolvable by the above means. All that is needed is that the bandwidth of the output filter be sufficiently reduced. One practical limitation here will be the bandwidth of the signal spectrum. Suppose, however, that the problem is simply one of determining the amplitude of a constant-amplitude sine wave and that the signal is available for as long as is required. Here the signal spectrum width is extremely narrow and again it would appear that there is no limit to how low the signal may become for it to be resolved.

The factor that places the limit in reality is PSD generated offset, drift and $1/f$ noise. This is quite distinct from the corresponding components generated by the signal amplifier. These can be rejected by suitable prefiltering. The PSD generated components, in contrast, cannot be so removed.

Suppose now that the output filter has been narrowed to the point where the converted input noise is insignificant compared with the PSD generated drift etc. The smallest input signal that can then be detected is that which produces an output comparable with PSD generated drift etc. The maximum input noise level is that which saturates the detector. Thus the minimum input signal-to-noise ratio for a resolvable signal is given by

$$(S/N)_{min} = \frac{\left(\begin{array}{c}\text{Input signal level giving output equal to}\\ \text{PSD-generated offset drift etc.}\end{array}\right)}{\text{Input noise level at which detector saturates}}. \qquad (7.16)$$

Dynamic range

It will now be shown that the above limiting signal-to-noise ratio at the PSD is virtually equal to D^{-1} where D is the 'dynamic range' of the PSD. D is defined as the ratio of the maximum to the minimum usable input signal levels for a noiseless input signal. Clearly the minimum signal level will be that to give an output matching the PSD generated drift etc, and the maximum input level will be the saturation signal input level. This is virtually identical to the saturation input noise level so the ratios are the same. Typical values for D are ~ 300 so the limiting input signal-to-noise ratio for a resolvable signal will be $\sim 300^{-1}$.

Where the input signal-to-noise ratio is too low to be resolved by these means, it is always possible to reduce the input noise level by narrow-band bandpass filtering of the signal at the PSD input. This is the only situation where *narrow*-band prefiltering is necessary, apart perhaps from the avoidance of strong interference components that might otherwise saturate the PSD. In all other cases the only prefiltering needed is the relatively broad-band type needed to remove unwanted acceptances. Any fine narrowing is done more conveniently by reducing the bandwidth of the low-pass filter at the PSD output.

8 Digitisation and noise

It is now very common to convert a required analogue signal into a series of binary numbers which are stored, processed and ultimately displayed by a small computer. The technique of multiple time averaging by computer described in chapter 2 is just one example of this. The process of converting the analogue signal into the binary (digital) numbers is called 'digitisation' or 'analogue-to-digital conversion'. This process, unless suitable precautions are taken, can increase the noise error for the measurement. Thus the main object of this chapter is to explain the noise increase and how it is avoided. In the process, the fundamental mode of operation of some of the more commonly used analogue-to-digital converter (ADC) circuits is described. Finally the process of 'aliasing' is discussed. This is a process whereby a number of different spectral components give the same set of digitised values. It is seen that, from the frequency-domain viewpoint, the mechanism whereby the above extra noise is admitted is that of aliasing.

8.1 ADDITIONAL NOISE DUE TO DIGITISATION

In this section the mechanism whereby digitisation can increase noise error is explained, together with the basic requirement for avoiding the increase. Figure 8.1(a) shows the usual arrangement when an analogue signal is digitised. Here, as usual, the noisy signal v_a is low-pass filtered to reduce the noise. This is essentially by the process of forming the running average of v_a and (b)–(d) illustrate the process in the usual way.

Then, at regular intervals of length T_s, the computer issues a command to the ADC to convert. For most types of ADC, the binary value output is essentially a

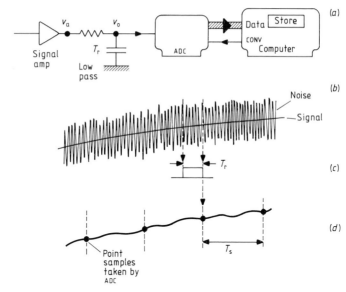

Figure 8.1 Normal method for taking and storing a series of digitised samples of an analogue signal. (*a*) Block diagram, (*b*) noisy signal v_a, (*c*) approximate memory function of filter, (*d*) filter output v_o.

point sample of the input. Thus the series of values eventually placed in the computer data store is that of the series of points shown in (*d*).

Thus each stored value actually represents the average of the noisy signal v_a over the period of length T_r immediately prior to the point sample. This means that, to avoid data wastage, T_r should equal T_s. For T_r/T_s the wastage factor is T_r/T_s, and this will cause an increase in white noise error by the factor $(T_r/T_s)^{-1/2}$.

Thus the means of noise increase is identified. To avoid the increase, the stored values should, by some means or other, be made the averages of the noisy signal v_a over the corresponding sampling interval of length T_s. Simply to take a point sample of v_a will give an increase by $(T_r/T_s)^{-1/2}$, where T_r is now the relatively short response time of whatever filter, internal to the signal amplifier, limits its frequency response.

Two minor points can be added. First, should T_r be made larger than T_s, this will waste no data and thus will cause no increase in noise error. However, the time resolution will be degraded unnecessarily, by merging together the input data over adjacent sampling intervals.

Second, the actual memory function for the CR filter is as shown in figure 8.2. The exponential taper is mildly undesirable, again having the tendency to merge adjacent sampling periods. This can be avoided by replacing the CR filter by the transversal running average filter shown in figure 8.3(*a*). Here, the n parallel outputs from the delay line are averaged using an analogue summing

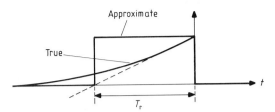

Figure 8.2 True memory function for the *CR* filter in figure 8.1(*a*), compared with the approximation made in figure 8.1(*c*).

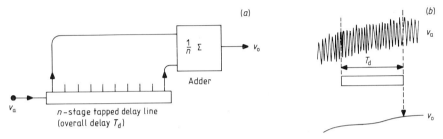

Figure 8.3 Use of transversal filter (*a*) to form true running average (*b*) of signal v_a to be digitised in figure 8.1.

circuit, scaled by the factor n^{-1}. Then, as shown in (*b*), the filter output is the running average of the input v_a over the period T_d of the delay line. Clearly, for our present purpose T_d should equal T_s. The advantages of this measure are here minor but the transversal running average filter will of further interest later.

8.2 ANALOGUE-TO-DIGITAL CONVERTERS

The purpose of this section is to describe the basic principle of operation of the three most commonly used analogue-to-digital converters. The precautions needed to prevent the noise increase of the last section are discussed in §8.3.

Digital-to-analogue converter

Figure 8.4 shows the elementary digital-to-analogue converter (DAC). This is an extremely simple device to describe, since the analogue output v_o is simply the analogue voltage corresponding to the digital input D. The response is more or less immediate (in fact in the region of 1 μs or less) and, unlike the ADC, there is no relatively complicated conversion sequence to be executed. The reason the DAC is described here is that it is frequently to be found as a key component in an ADC, incorporated into some sort of feedback circuit.

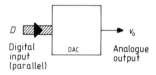

Figure 8.4 Digital-to-analogue converter.

Counting ADC

The counting ADC system shown in figure 8.5 is one such system. Here the DAC is driven by an oscillator and counter so that, after reset, the DAC output voltage v_d describes the 'staircase' waveform shown. v_d is compared with the analogue input voltage v_{in} and, when v_d exceeds v_{in}, the comparator output $\overline{\text{HI}} \rightarrow 0$, which resets the bistable and stops the counter. At this point, since v_d is as close to v_{in} as the DAC resolution allows, the counter output is the digital representation of v_{in}.

$\overline{\text{CONV}}$ is the command pulse from the computer, which initialises the process. The bistable output provides a convenient BUSY signal, which indicates to the computer when the conversion is complete and the digital output may be read.

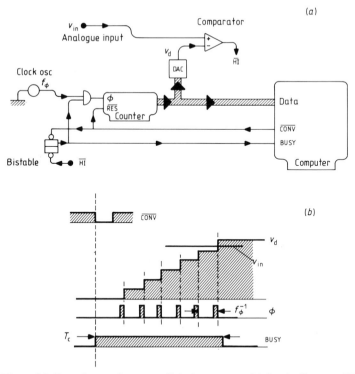

Figure 8.5 Counting analogue-to-digital converter. (*a*) Logic diagram, (*b*) waveforms.

The conversion time T_c for this device depends on the clock frequency f_ϕ and the resolution of the converter. For an eight-bit converter and $f_\phi = 1$ MHz, the maximum value of $T_c = 1\ \mu s \times 2^8 = 256\ \mu s$.

Successive approximation ADC

The successive approximation ADC in figure 8.6 is a refinement of the counting ADC which reduces the conversion time. The counter, clock oscillator etc in the counting ADC are replaced by the successive approximation register, which is controlled by the program of figure 8.6(b). Initially all bits are set to zero. Then the most significant bit is set. If the resulting $v_d > v_{in}$ the bit is reset, otherwise not. Then the next most significant bit is set, and so on, down to the least significant bit. For an eight-bit converter, this gives a conversion time $T_c = 8 \times f_\phi^{-1} = 8\ \mu s$ for a 1 MHz clock oscillator. This contrasts with $T_c = 2^8 f_\phi^{-1}$ for the counting ADC. Clearly the disparity between the two values increases with the resolution time of the converter.

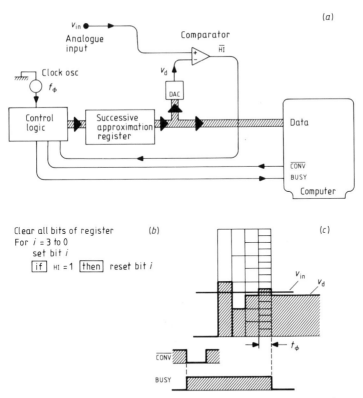

Figure 8.6 Successive approximation analogue-to-digital converter (four-bit for clarity). (a) Block diagram, (b) controller program, (c) waveforms.

The two methods of conversion are sometimes likened to the process of weighing an object in a pair of scales. Here the object to be weighed is v_{in} and the balance is the comparator. The counting ADC is equivalent to adding a series of small weights of the same value (say 1 gram) to the scales, until they tip for the first time. The eight-bit successive approximation converter, in contrast, uses a set of graded weights $(2^7 g, 2^6 g, \ldots, 1 g)$ in the more usual manner.

Dual ramp ADC

Figure 8.7 shows the dual ramp ADC, which is based on an analogue integrator. After phase (1) in the controller routine of (c) the integrator output v_x is given by

$$v_x = -CR^{-1} \int_0^{T_0} v_{in} \, dt. \qquad (8.1)$$

Or if, for simplicity, it is assumed that v_{in} is constant,

$$v_x = -T_0 v_{in}/CR. \qquad (8.2)$$

During phase (2), v_x becomes zero after a time T_1 such that

$$v_x = T_1 v_{ref}/CR \qquad (8.3)$$

where v_x is as given in equation (8.2). Thus, from (8.2) and (8.3),

$$T_1 = v_{in} T_0/v_{ref}. \qquad (8.4)$$

But T_0 and v_{ref} are constant, so $T_1 \propto v_{in}$. Further, at the end of phase (2), when the counter is disabled, the count will be proportional to T_1, and therefore to v_{in}. Thus the counter output becomes the ADC output.

When v_{in} is not constant, equation (8.1), rather than (8.2), should be combined with (8.3), giving

$$T_1 = v_{ref}^{-1} \int_0^{T_0} v_{in} \, dt. \qquad (8.5)$$

Thus the dual ramp converter produces an average of v_{in} over T_0, rather than a point sample.

The dual ramp converter is capable of very high accuracy at moderate cost, largely because it does not include a DAC. For the DAC, accuracy commensurate with the resolution is relatively easy to obtain for an eight-bit converter but becomes difficult for higher values, sixteen bits being probably the limit. However, for 16-bit accuracy and more, the conversion time T_c for the dual ramp device becomes large. For example, for 16-bit resolution and $T_\phi = 1 \mu s$, $T_c = 1 \mu s \times 2^{16} \simeq 64$ ms.

A common practice is to make the period T_0 over which the signal is integrated equal to 20 ms. This is the period corresponding to one cycle of 50 Hz interference and completely eliminates the effect.

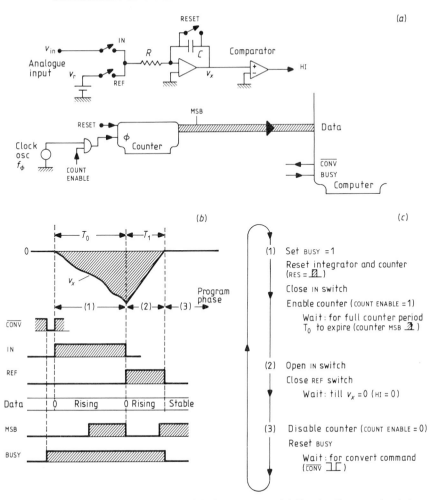

Figure 8.7 Dual-ramp analogue-to-digital converter. (*a*) Logic diagram (omitting controller), (*b*) waveforms, (*c*) controller program.

8.3 SUITABILITY OF REAL ADC CIRCUITS FOR TAKING AVERAGE SAMPLES

We now consider how the digitisation processes used in the three types of ADC described in the last section allow them to be used to produce samples which represent the averages of the analogue input signal over the sampling period as in figure 8.1.

The counting type ADC shown in figure 8.5 produces a point sample, being the value of the input v_{in} at the time at which v_d passes v_{in}. Thus, in principle, the device could be used as the point sampling ADC in figure 8.1, backed by the necessary prefilter to form the averages. However, the delay between the command signal and the sampling point is proportional to v_{in}. Thus, the sampling intervals will no longer be quite regular. This will cause the averaging periods of length T_s (the interval between the command pulses) to overlap in some cases and in others to form gaps in which data are wasted. The noise increase will be small in all cases and certainly so when the conversion time T_c is small compared with T_s.

The successive-approximation converter of figure 8.6 does not produce a single point sample. Normally it is expected that the input v_{in} will be constant during the conversion period T_c. If it is not, the value output is always between the maximum and minimum values of v_{in} over the period. Thus, again, the device can be used as the point sampling ADC in figure 8.1, totally satisfactorily if $T_c \ll T_s$ and with little noise increase even when T_c approaches T_s.

The dual ramp converter of figure 8.7 appears to be ideally suited to the present application. Its samples are formed at regular intervals and they are actually averages of v_{in}, over the period T_0. Thus, even if the prefilter is omitted, the data wastage factor will be finite, i.e. T_0/T_s.

Inclusion of the filter will then make good even this loss, extending the averaging period to T_s as required. However the dual ramp converter, although extremely accurate, is very slow in operation compared with the successive approximation ADC and this will often preclude its use.

8.4 POINT SAMPLING GATE

Figure 8.8(a) shows an arrangement that overcomes the timing uncertainties of both the counting and the successive-approximation-type ADC circuits. Here the ADC is preceded by a 'track-and-hold' circuit. Suitably operated, this can be made to take a point sample of the input v_{in} and to hold it for as long as the ADC is converting.

To understand the operation of the circuit, consider the track-and-hold function illustrated in figure 8.9. Here, when the switch is closed, $v_h = v_{in}$, i.e. v_h 'tracks' v_{in}, while when the switch is open, v_h 'holds' at the value of v_{in} at the time of opening.

If, therefore, the computer closes the switch prior to the sampling time and opens it at the sampling time, as in figure 8.8(b), then the point sample is held for a time long enough to allow conversion.

In principle, the time T_t for which the track switch needs to be closed can be made vanishingly small. In this instance the gate is usually referred to as a 'point-sampling' gate, or 'sample-and-hold' gate, rather than a 'track-and-hold'. However, it must be emphasised that the value held by each is, in principle, a true point sample, regardless of how long T_t might be.

In practice T_t must be large compared with $R_s C$, where R_s is the signal source

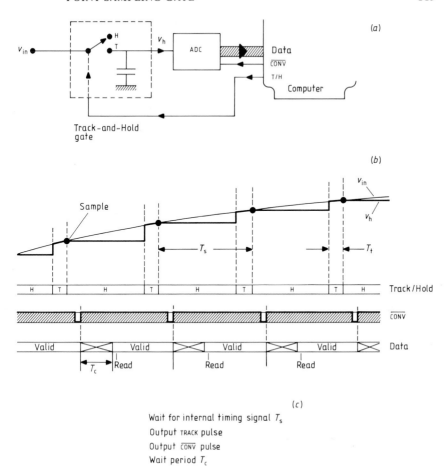

Figure 8.8 Adaptation of normal ADC to give point sampling at regular intervals. (a) Block diagram, (b) waveforms, (c) computer program for each sampling period T_s.

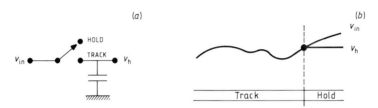

Figure 8.9 Track-and-hold gate. (a) Diagram, (b) waveforms.

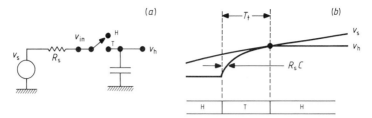

Figure 8.10 Diagram showing the need for $T_t \gg R_s C$ when using the track-and-hold gate. (*a*) Diagram, (*b*) waveforms.

impedance, as shown in figure 8.10. With R_s finite, the 'point samples' then actually represent averages over the period CR_s. When, as is likely, the switch is a FET, the channel impedance will also contribute to R_s.

Point-by-point differentiation

From the noise viewpoint, the inclusion of the point-sampling gate gives small return. However, there is one application where the device is more necessary. Sometimes the differential of the input signal is required, expressed as the difference Δv_{in} between successive samples. Clearly, $dv_{in}/dt = \Delta v_{in}/T_s$ and, since T_s is constant, dv_{in}/dt can be taken as being proportional to Δv_{in}. However, when the conversion time T_c approaches T_s, the scatter in T_s approaches 100% and the error in dv_{in}/dt becomes gross. Inclusion of the sampling gate overcomes this problem.

Analogue multiplexing

Another instance when the point sampling gate is used is shown in figure 8.11. Here the single ADC is multiplexed to digitise a series of input channels sampled simultaneously. This is merely an economic measure to save the cost of a number of ADCs.

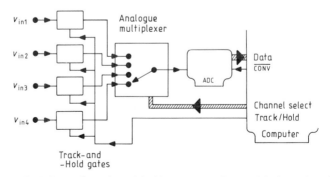

Figure 8.11 Use of track-and-hold gates to give multi-channel point sampling with a single ADC.

8.5 ALIASING

The problem of aliasing is introduced in figure 8.12. Here f_1 is the frequency of
the required signal, which is sampled at the frequency f_s. It is then seen that the
interference component of frequency $2f_s + f_1$ produces the same set of samples.
Actually, all of the components in figure 8.13(a) produce this set of samples,
which we will now show.

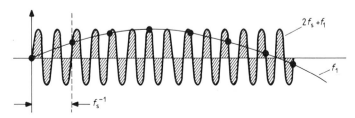

Figure 8.12 Demonstration that sine waves of different frequency can
produce the same set of samples.

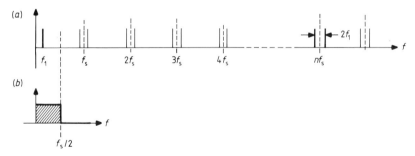

Figure 8.13(a) The aliases associated with a sine wave of frequency f_1 sampled at the
frequency f_s. (b) Frequency response of anti-aliasing filter.

We shall let $\cos(2\pi f_1 t)$ represent the component of frequency f_1 and
$\cos[2\pi(nf_s + f_1)t]$ the upper component shown emphasised. Then at the mth
sampling point $t = mT_s$ and the component of frequency f_1 is $\cos(2\pi f_1 mT_s)$ and
that of frequency $nf_s + f_1$ is $\cos(2\pi nm + 2\pi f_1 mT_s)$. But, since m and n are integers,
these are the same. Also, for the component of frequency $nf_s - f_1$ shown
emphasised, the value is $\cos(2\pi nm - 2\pi f_1 T_s)$, which is again equal to the
component of frequency f_1. Thus, each of the three emphasised components
has the same samples and, since n can have any integer value, all of the
components shown in figure 8.13(a) have the same samples.

The difficulty is that, without sampling, it is easy for the processor to
distinguish the high-frequency interference components from the signal,
simply from their frequencies, but once sampled, and with the aliasing effect
making the samples the same, this distinction can no longer be made.

As the value of f_1 in figure 8.13(a) is increased, all aliases are of higher frequency than f_1 for as long as $f_1 < f_s/2$. Thus, provided the signal frequency f_1 is kept within this range, the anti-aliasing filter response shown in figure 8.13(b), applied at the input of the sampling device, will remove all aliases.

Spectral analysis

Figures 8.14(a) and (b) compare the frequency response of the filter providing the running average in figure 8.1 with the present anti-aliasing filter. The two are very similar, both having a cut-off frequency comparable with the sampling frequency $f_s = T_s^{-1}$, and it is clear that there is a strong connection between aliasing and the additional noise error obtained by omitting the filter. This connection is made clearer as we now apply the methods of analysis used for the PSD circuits in chapter 7 to obtain the spectral response of the process of digitisation.

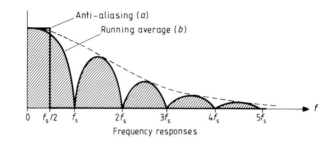

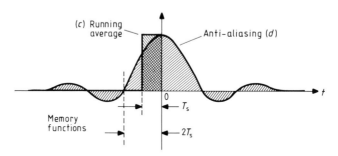

Figure 8.14 Comparison of frequency responses and memory functions for (a), (d) the anti-aliasing filter and (b), (c) the running average filter of figure 8.3.

The basic mechanism of forming a series of point samples of the input waveform v_a is represented by the analogue multiplier shown in figure 8.15(a). The 'reference' input v_r to this PSD-like device is then the periodic train of impulses shown in (c) and the resulting output v_p becomes the weighted impulse train shown in (d). Here the magnitude of each output impulse is

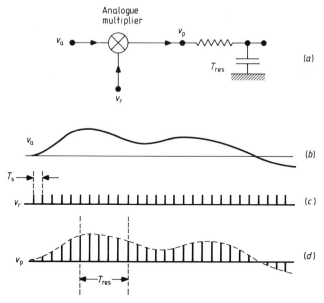

Figure 8.15 Circuit model (a) for sampling and reconstruction of signal v_a.

proportional to the value of the corresponding point sample and so the output train is an adequate representation of the series of stored samples.

Figure 8.16 shows the spectral view of the process so far. As in chapter 7, we start by assuming the input to be a single spectral component of frequency f_a. The reader should be warned that some change of notation has been necessary; the subscript 's' now implies 'sampling' not 'signal' as in chapter 7.

Next, (b) shows the 'reference' spectrum of the regular sampling impulse train v_r. This gives the normal 'sum and difference' output spectra of (c).

Suppose next that the final objective is to evaluate the required signal over one of the time-resolution periods of length T_{res}, as shown in figure 8.15(d). This is equivalent to subjecting the output pulse train v_p to a low-pass filter of cut-off frequency $(2\pi T_{res})^{-1}$. Assuming $T_{res} > T_s$, the cut-off will be well below the sampling frequency f_s, as shown in (d). Thus, the effective acceptance pattern becomes as in (e).

Finally (f) shows the frequency response of the anti-aliasing filter required at the input of the digitiser. This removes all but the lowest acceptance and converts the frequency response to that in (d), which is the same as that obtained when, without digitisation, the noisy analogue signal v_a is averaged over the period T_{res}. The final white noise error will thus be the same as for this simple process.

If the input acceptance pattern in (e) is compared with the series of aliases in figure 8.13(a), it is clear that the noise components in the upper acceptances are the aliases of those in the lowest acceptance. Thus the low-pass anti-aliasing

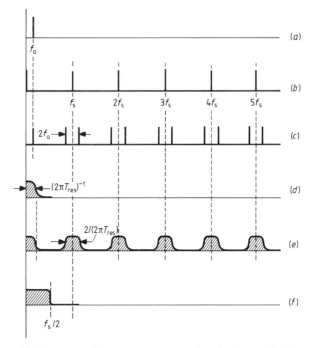

Figure 8.16 Spectra and frequency responses for circuit model of figure 8.15. (a) Spectrum of input signal v_a, (b) spectrum of sampling pulse train v_r, (c) spectrum of v_p, (d) frequency response of output filter corresponding to averaging over T_{res}, (e) input acceptance pattern, (f) frequency response of anti-aliasing filter.

filter avoids extra noise when digitising by rejecting those components that are aliases of those normally accepted when the signal is averaged over T_{res} without digitisation.

Previously the function of the low-pass filter preceding the ADC was considered to be that of avoiding additional noise error by extending the noise averaging period to the whole of each sampling period. The above is therefore the frequency-domain view, and this the time-domain view, of what is essentially the same process.

Whichever view is taken, the loss is irretrievable if the filter is absent or has too high a cut-off frequency (too low an averaging period). From the time-domain viewpoint, no subsequent manipulation of the stored samples can recover the lost data. From the frequency-domain point of view, the irretrievability lies in that the aliased noise components become spectrally indistinguishable from the legitimate components in the lowest acceptance and so cannot subsequently be removed by filtering the stored samples.

From figure 8.14, the running average and the anti-aliasing filters are not quite identical and the question therefore arises as to which should be used in any given situation. If the running average filter is used, the noise performance is totally satisfactory, but the removal of aliases above $f_s/2$ will not be complete, even though aliases at exact multiples of f_s are completely cancelled by the filter zeros.

If, instead, the anti-aliasing filter is used, the rejection of aliases will be total. Figures 8.14(c) and (d) compare the memory functions for the two filters. The width of the anti-aliasing function is seen to be approximately $2T_s$. This is more than enough to avoid data wastage and so the noise error is again not increased. However there is a significant loss in time resolution which will require doubling the sampling rate and also the amount of data storage memory. It therefore appears that, when noise alone is the limiting factor, the running average filter is adequate, but, when there are strong high-frequency interference components present, the anti-aliasing filter will be necessary, with the additional required storage space.

9 Magnitude determination for transient signals of known shape and timing

Up to the present we have been concerned with the measurement of signals that are continuously present. This is not always so and one example is the measurement of a continuous pulse train. Here the pulse shape and timing is often known and it is required to determine any progressive change in amplitude of the pulses. In this chapter we consider how to do this with minimum error due to noise, drift etc. The treatment extends to single pulses and to transients of more complex shape. The problem of measuring the maximum and minimum value of a signal is amenable to the same kind of analysis and is also included. The associated problem of measuring the time of occurrence of such a signal transient is covered in the next chapter.

9.1 FLASH SPECTROMETER

The arrangement of figure 9.1 constitutes an example which will be used to develop the general approach. The purpose of the flash spectrometer is to measure the optical spectrum from the photoflash tube, a device which of necessity is pulsed. As the flash is periodically triggered, the wavelength λ of the monochromator is scanned linearly over the required range. (A monochromator is a device which transmits light of essentially one wavelength λ, which is variable. It is the optical equivalent of a tunable narrow-band bandpass filter.) Thus the amplitude of the train of output pulses v_a is indicative of the tube spectrum.

The method used to record the pulse amplitudes will depend upon whether digitisation of the amplitudes is required or the result is to be output directly to a chart recorder. In the absence of any kind of unwanted signal, digitisation is

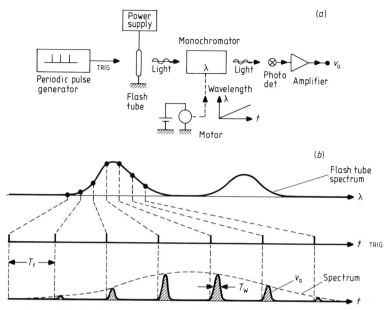

Figure 9.1 Flash spectrometer. (*a*) Block diagrams, (*b*) waveforms.

achieved by triggering the ADC at the peak of each pulse. This requires that the conversion time T_c of the ADC is small compared with the width T_w of the signal pulse. Where this is not so, inclusion of a point-sampling gate before the converter, as in chapter 8, resolves the problem.

When direct chart recorder display is required, all that is needed is the point-sampling gate between the spectrometer output and the recorder, again with the gate triggered at the peak of each pulse. In this way the peak height is held from one peak to the next.

Transversal running average filter

When white noise is present, point-sampling is not acceptable and some sort of average of the signal is required. Figure 9.2 shows one approach. Here the transversal running average filter and point-sampling gate are used as in figure 8.3, but this time with the averaging period T_d equal to the signal pulse width T_w. The arrangement is for digitisation but, as above, if direct chart recorder display is required, the recorder can follow the sampling gate directly.

Integrate-and-hold circuit

Figure 9.3 shows an alternative method which uses an analogue integrator in place of the transversal filter and sampling gate. Another way of avoiding the need to use the expensive transversal filter is, as in the last chapter, to replace it by a simple CR low-pass filter. Here CR should equal the signal pulse width, as shown in figure 9.4.

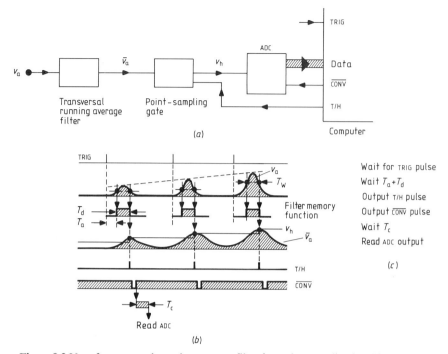

Figure 9.2 Use of transversal running average filter for pulse sampling in white noise. (*a*) Block diagram, (*b*) waveforms, (*c*) computer program.

Low-pass filter

A very simple method of processing the pulse train is to pass it through a low-pass filter of cut-off frequency below f_f, the frequency of the pulse train. The filter output can then be digitised or applied directly to the chart recorder. This method tends to 'smear' the pulses; in other words, it produces a recording which represents the mean of the last few pulses, rather than resolving the separate amplitude of each pulse. This is sometimes a disadvantage but not really so for the example of the flash spectrometer.

Probably the most serious objection to the low-pass filter method lies in the way that it exaggerates the effects of offset and drift. For a signal pulse width T_W and a period T_f between pulses, the magnification is by T_f/T_W.

9.2 LIMIT SETTING

For the running average and the integrate-and-hold systems of figures 9.2 and 9.3, the limits t_1 and t_2 to the period over which the noisy signal pulse is accepted are shown as being the times when the signal is at half height. This is

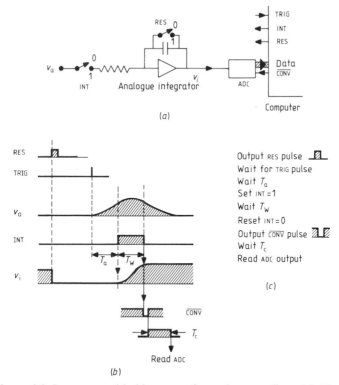

(a)

Output RES pulse
Wait for TRIG pulse
Wait T_a
Set INT =1
Wait T_w
Reset INT=0
Output $\overline{CONV}$ pulse
Wait T_c
Read ADC output

(c)

(b)

Figure 9.3 Integrate-and-hold system for pulse sampling. (a) Block diagram, (b) waveforms, (c) computer program.

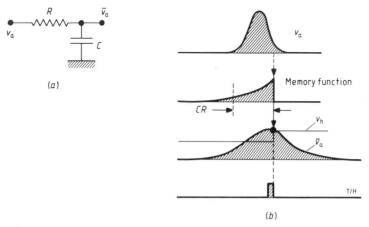

(a)

(b)

Figure 9.4 Use of simple CR filter in place of transversal filter of figure 9.2. (a) Filter, (b) waveforms.

approximately correct and in this section the exact values of t_1 and t_2 for minimum noise error are calculated.

Rectangular pulse

Consider first a rectangular pulse and compare the cases when, for a pulse-width T_W and an averaging period T_{av}, $T_{av} \ll T_W$ and $T_{av} \gg T_W$. When white noise is averaged over the period T_{av}, the standard deviation of the noise average is proportional to $T_{av}^{-1/2}$. For $T_{av} \ll T_W$ the signal average is independent of T_W, so the signal-to-noise ratio $S/N \propto T_{av}^{1/2}$. Conversely, when $T_{av} \gg T_W$ the signal average $\propto T_{av}^{-1}$ so $S/N \propto T_{av}^{-1/2}$. Thus S/N is maximum when $T_{av} \simeq T_W$.

This is a further condemnation of the method of averaging the pulses with a low-pass filter of cut-off frequency below the pulse repetition rate. If T_f is the period between pulses, the signal-to-noise ratio is now seen to be degraded by the factor $(T_f/T_W)^{1/2}$ when the low-pass filter is used instead of averaging over the pulse width only.

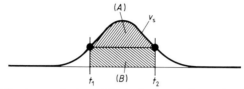

Figure 9.5 Optimum placing of limits for averaging a signal pulse in the presence of white noise (area (A) = area (B)).

Non-rectangular pulse

For a non-rectangular pulse such as that shown in figure 9.5 the position is more complex. In fact the limits have to be such that the areas (A) and (B) shown are equal. To demonstrate this point, let σ_i represent the standard deviation for the noise integrated over the period from t_1 to t_2 and I_s the integral of the signal component v_s of the noisy signal v_a over the same period. Then, since $\sigma_i = k(t_1 - t_2)^{1/2}$,

$$\sigma_i/I_s = k(t_2 - t_1)^{1/2} \bigg/ \int_{t_1}^{t_2} v_s \, dt. \tag{9.1}$$

To find the best value for the first end-point t_1 of the integral, differentiate σ_i/I_s with respect to t_1 and set the result to zero. This gives the condition

$$v_s(t_1) \cdot (t_2 - t_1) = \tfrac{1}{2} \int_{t_1}^{t_2} v_s \, dt. \tag{9.2a}$$

In the same way, the best value for t_2 is given by

$$v_s(t_2) \cdot (t_2 - t_1) = \tfrac{1}{2} \int_{t_1}^{t_2} v_s \, dt. \tag{9.2b}$$

But area (B) in figure 9.5 is equal to the LHS of both of these equations and $\int_{t_1}^{t_2} v_s \, dt$ in the RHS $= (A) + (B)$. We thus require $(B) = [(A) + (B)]/2$, i.e. $(A) = (B)$ as proposed.

9.3 WEIGHTED PULSE SAMPLING

The above averaging of a noisy signal pulse between the optimum limits t_1 and t_2 can be envisaged as being carried out by the circuit shown in figure 9.6(a). Here the integrator operates continuously and the weighting function w in (c) limits the integral to the required period. The question then arises as to whether the noise error can be reduced still further by grading the weighting function as in (d). In fact, some further improvement can so be obtained and it

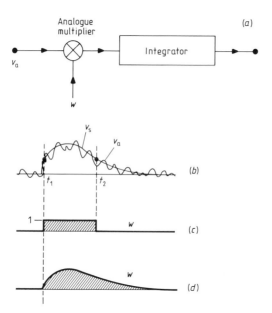

Figure 9.6 Circuit representation of average formation in figure 9.5. (a) Block diagram, (b) signal, (c) weighting function for figure 9.5, (d) replacement of abrupt weighting function in (c) by graded weighting, proportional to v_s, giving best possible signal-to-noise ratio.

is the main object of this section to establish the important general result that, for white noise, the noise error is minimum when the shape of the weighting function w is the same as that for the required signal v_s, i.e. if

$$w = k_w v_s \tag{9.3}$$

where k_w is some constant, the value of which is immaterial. This strategy is only applicable, of course, if the shape and timing of the signal is known and the factor to be measured is its amplitude.

To justify the above statement, we first divide the signal pulse into a large number of narrow intervals of width Δt. Let σ_Δ represent the standard deviation of the noise integral over each of these periods. Also let σ_w be the standard deviation for the weighted integral of the noise over the whole signal pulse. Then, because the values of the noise integrals for the various integrals are uncorrelated,

$$\sigma_w = \left(\sum_i w_i^2 \sigma_\Delta^2 \right)^{1/2} \tag{9.4}$$

where w_i is the value of w for the ith interval.

Let I_w represent the weighted integral $\int_{-\infty}^{+\infty} v_s w \, dt$ of the signal pulse. Then

$$I_w \simeq \sum_i v_{si} w_i \, \Delta t \tag{9.5}$$

where v_{si} is the value of v_s over the ith interval.

The expected fractional error σ_w / I_w in the estimation of I_w is then given by

$$\sigma_w / I_w = \left(\sum_i w_i^2 \sigma_\Delta^2 \right)^{1/2} \Big/ \sum_i v_{si} w_i \, \Delta t. \tag{9.6}$$

If now σ_w / I_w is differentiated with respect to any one of the weights, say w_j, it is found that, if $w = k_w v_s$, then $d(\sigma_w / I_w)/dw_j = 0$. This means that if any one of the weights is changed from the value giving the condition $w = k_w v_s$, σ_w / I_w is increased. Thus it is established that, for the ratio σ_w / I_w of noise error to signal to be lowest, the weighting function w must have the same shape as the required signal transient v_s.

The minimum value $(\sigma_w / I_w)_{min}$ of σ_w / I_w is given by inserting the condition $w = k_m v_s$ into equation (9.6). This gives

$$\left(\frac{\sigma_w}{I_w} \right)_{min} = \frac{\sigma_\Delta / \Delta t^{1/2}}{(\sum_i v_{si}^2 \, \Delta t)^{1/2}}. \tag{9.7}$$

Now the noise error $\sigma_\Delta \propto \Delta t^{1/2}$, being that for an integral, i.e. $\sigma_\Delta = k \, \Delta t^{1/2}$. Using this expression and allowing $t \to 0$, equation (9.7) becomes

$$\left(\frac{\sigma_w}{I_w} \right)_{min} = \frac{k}{(\int_{-\infty}^{+\infty} v_s^2 \, dt)^{1/2}}. \tag{9.8}$$

Integrate-and-hold realisation

Figure 9.7 shows one way of realising the above optimally weighted pulse sampling. This is an almost direct translation of the block diagram of figure 9.6 into analogue hardware. It can also be thought of as an extension to the simple

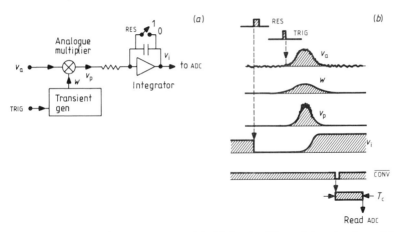

Figure 9.7 Adaptation of the integrate-and-hold circuit of figure 9.4 to give the weighted integration of figure 9.6(a) and (d). (a) Block diagram, (b) waveforms.

integrate-and-hold system of figure 9.4. Here the translation from the rectangular weighting function of figure 9.6(a) to the ideal weighting of (d) is achieved by replacing the INT switch in the simple circuit by the analogue multiplier and transient generator of figure 9.7.

Transversal filter realisation
It is equally possible to adapt the transversal filter system of figure 9.2 to give weighted pulse sampling. The filter, initially shown in figure 8.3, produces a running average of the input signal. This requires that all the inputs to the analogue adder have equal weight. If, however, the weights can be made unequal, it is possible to make a filter with any kind of memory function. Figure 9.8(a) thus shows the filter with the weights proportional to the required signal, which is the optimum weighting.

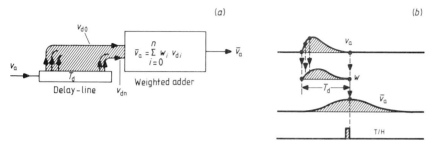

Figure 9.8 Modification of transversal filter of figure 9.2 to give the optimally weighted integration of figure 9.6(d) (matched filtering). (a) Block diagram, (b) waveforms.

Notice that the filter imposes a delay T_d and this requires the point sample of the filter output to be taken a suitable period after the peak of the signal pulse. Once again, if the conversion time of the ADC is small compared with the signal pulse width, the point sampler can be omitted.

PSD **realisation**

Figure 9.9 shows the integrator in figure 9.7 replaced by a CR filter. If the time constant T_r of the filter is large compared with the period T_f of the signal pulse train, the filter output v_o will represent the running average of the last T_r/T_f output pulses from the analogue multiplier. Here the optimally weighted pulse sampling is realised somewhat more simply than by the weighted integrate-and-hold system, but at the expense of the individual resolution of each pulse.

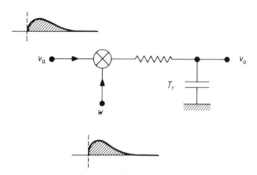

Figure 9.9 PSD realisation of optimally weighted pulse sampling of figure 9.6(d).

When the system of figure 9.9 is compared with the PSD systems of figure 7.4 it is found to be identical in layout. The analogue multiplier in figure 9.9 simply constitutes a PSD for which the reference waveform v_r is the same in shape as the signal waveform. The analogue multiplier PSD in figure 7.4(α) and the reversing PSD of figure 7.4(β) can now be viewed as particular examples of this general principle. For the sine wave input signal and analogue multiplier PSD the best reference waveform is a sine wave, while for a square wave signal the best reference waveform is a square wave, provided most simply by the reversing PSD.

Now, for a pulsed signal the best reference is a pulse train of the same shape. A good approximation is afforded by a train of rectangular pulses of width equal to the signal pulse width T_w. Note in particular that if the reversing PSD is used here, with a square wave reference, in ignorance of this principle, the loss in signal-to-noise ratio is of the order of $(T_w/T_f)^{1/2}$.

More complex signal transients

When the signal transient is more complex than the simple pulse so far considered, accommodation is straightforward for the weighted transversal filter system of figure 9.8 and the weighted integrate-and-hold system of figure 9.7. All that is needed is to alter the filter memory function or the transient generator to suit, so that the weighting function has the shape of the new signal. The problem becomes more difficult, however, in the cheaper approximations to these ideals. Figure 9.10 shows a way of modifying the simple integrate-and-hold circuit of figure 9.4 to suit a doublet signal. Without such modification the gate would give zero response to such a signal, since its integral is zero, unless the switch was only closed for one half of the signal period.

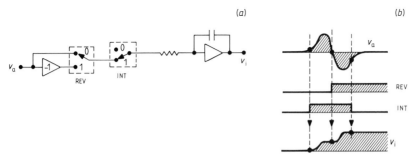

Figure 9.10 Modification to simple integrate-and-hold system of figure 9.4 to allow for antisymmetric signal transient (doublet signal). (*a*) Block diagram, (*b*) waveforms.

Alternatively one could use the simple CR filter and sampling gate of figure 9.4, but taking two samples as in figure 9.11. These are subsequently subtracted to give the effective memory function shown.

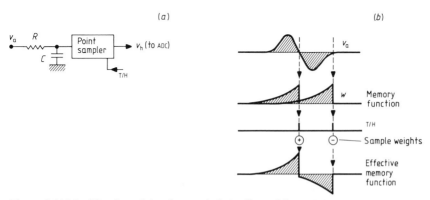

Figure 9.11 Modification of simple sampled CR filter of figure 9.4 to accommodate an antisymmetric signal transient. (*a*) Block diagram, (*b*) waveforms.

The extension of these methods to yet more complex signal shapes is obvious, by suitably increasing the number of times the REV switch is operated for the integrate-and-hold, or taking a suitably increased number of samples for the sampling gate system.

A further alternative for the simple doublet signal of figure 9.10 is to use the more complex CR filter of figure 10.15. This is based on an op-amp differentiator circuit and gives a weighting function that is moderately well matched to the doublet shape. Only single sampling is then required.

9.4 MATCHED FILTERING

The weighted transversal filter in figure 9.8 is an example of a 'matched filter'. Here the matching is between the memory function of the filter and the waveform of the required signal transient v_s. This is a time-domain view of what is normally regarded as a frequency-domain process. It will now be shown that the corresponding frequency domain 'matching' criteria are (a) that the filter gain-frequency response should match the amplitude-frequency function for the signal spectrum and (b) that the phase-frequency response for the filter should be the negative of that of the phase-frequency function for the signal spectrum. As before, these are the requirements for minimum white noise error.

The argument is developed using figure 9.12. Here F_s^* in (b) represents the required transversal filter. The point sampler then represents either the usual

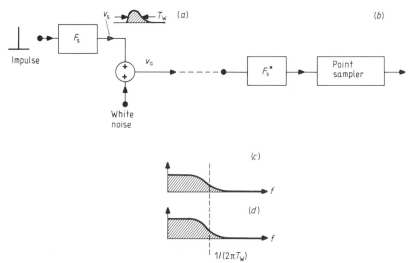

Figure 9.12 Spectral view of the matched filter pulse sampling system of figure 9.8. (a) Synthesis of noisy signal, (b) transversal filter of figure 9.8, (c) spectrum of signal v_s, (d) gain-frequency response of filter F_s^*, $F_s = F_s^*$, $\phi(F_s^*) = -\phi(F_s)$.

ADC or the point sampling gate. The noisy signal v_a is synthesised using the filter F_s which has an impulse response equal to v_s. This is then mixed with white noise.

The matching of the filter F_s^* to the signal means that the memory function of F_s^* matches the waveform of v_s. But the memory function of F_s^* is its impulse response reversed in time, and v_s is the impulse response of F_s, so it follows that the impulse response of F_s^* is the time-reversed impulse response of F_s.

Suppose, now, that when the impulse is applied to F_s one of the spectral output components is given by $A \cos(\omega t + \phi)$. The corresponding component when the impulse is applied to F_s^* will be the time reversal of this, i.e. $A \cos(\omega(-t) + \phi)$ and this is equal to $A \cos(\omega t - \phi)$.

For both filters the amplitude A is the same, so the amplitude-frequency response for F_s^* is equal to that for F_s. Also, for the input impulse, the phase at $t = 0$ is zero for all frequencies. Thus the phase-frequency response for F_s^* is the negative of that for F_s.

Again, since for the input impulse in figure 9.12(a) all input components are of the same amplitude and all phase angles are zero, it follows that the spectrum for the signal v_s is equal to the frequency response of F_s. This means that the amplitude-frequency response of F_s^* is equal to the amplitude-frequency function for the signal and the phase-frequency response for F_s^* is the negative of the phase-frequency function for the signal. Thus the requirements (a) and (b) above are confirmed.

The matching of filter frequency response to signal spectrum amplitude is a widely acknowledged aspect of matched filtering. Here, the frequency domain is divided into a large number of intervals of differing signal-to-noise ratio. The matched filtering is then the process of emphasising those intervals for which the signal-to-noise ratio is high and suppressing those for which it is low. This is the obvious frequency-domain counterpart of dividing the time scale into a similar series of intervals and again weighting those for which the signal-to-noise ratio is high more favourably than those for which it is low.

The associated requirement for the phase-frequency response of the filter is less widely acknowledged and less obvious in purpose. Starting at the input impulse in figure 9.12, all of the spectral components are in phase at $t = 0$. The filter F_s then dephases them to give the distribution for the real input signal v_s. The phase shift introduced by F_s^* then cancels that due to F_s and brings all the components into phase at $t = 0$. This gives the filtered signal the maximum value possible and, incidentally, makes $t = 0$ the correct time to sample the filter output. For other than this phasing, the maximum signal level will, in general, be less and so the signal-to-noise ratio inferior.

A common error when unaware of the phase requirement is to make F_s^* identical to F_s. This has the effect of making the memory function the reverse of what it should be. Figure 9.13 illustrates the effect. Here, the required signal v_s has the simple exponential shape resulting from the impulse response of the CR filter in (a). The filter (b) has a gain-frequency response matching the signal

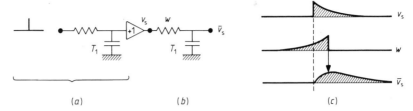

Figure 9.13 Erroneous attempt at 'matched filtering' using a filter which has the same gain-frequency response as the signal v_s but without the necessary phase reversal (time-reversal of the impulse response). (a) Signal simulation, (b) 'matched' filter, (c) waveforms.

and thus is assumed adequate. Then if the filter output v_s is sampled at its maximum value, this is thought to constitute the ideal matched filtering.

Comparison of memory function and v_s in the diagram reveal that no such matching exists and this is because property (b) above has been ignored. Failure to include the phase reversal implies omission of the corresponding time reversal, giving the disparity shown.

This is hardly an error we should be likely to fall into, having developed the concept of matched filtering initially in the time domain.

One final point should be noted. In fact, the impulse response of F_s^* in figure 9.12 cannot be realised, because it anticipates the input impulse. What can be realised, however, is a filter with an impulse response of the required shape, but delayed in time, to follow the impulse rather than precede it. This will simply require sampling to be delayed, as in the real system of figure 9.8.

PSD

The spectral response of the PSD systems of figure 7.4 confirm the above principles. Notice that both for the sine wave signal applied to the analogue multiplier with the sine wave reference and for the square wave signal with the effective square wave reference of the reversing PSD, not only is the reference always of the same shape as the signal, but also the spectral acceptance pattern of each PSD matches the corresponding input spectrum.

9.5 BOXCAR SAMPLING GATE

It has been shown that the optimally weighted pulse sampling afforded by the modified integrate-and-hold and transversal filter circuits of figures 9.7 and 9.8 gives an improvement in noise error relative to the equivalent unmodified circuits of figures 9.2 and 9.3. However, it will be clear that for an ordinary pulse, e.g. of the type produced by the flash spectrometer, the improvement is modest. Thus, it is fairly likely that one of the unweighted systems will be used in practice.

Of these, the running-average transversal filter is fairly expensive and both are fairly slow in operation, the transversal filter because it is based on a CCD delay line and the integrate-and-hold circuit because of the analogue integrator. These factors make the very simple 'boxcar' circuit of figure 9.14 an attractive alternative. This, it will be shown, produces true unweighted averaging sampling between limits, as for the other two, and has only the minor disadvantage of giving an output that represents the average of the last few pulse samples, rather than resolving each pulse separately.

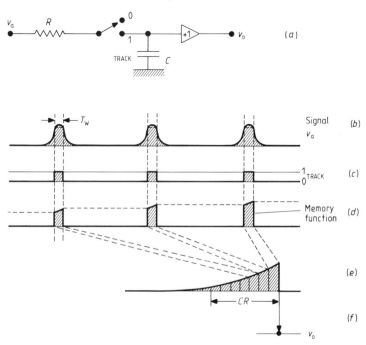

Figure 9.14 Boxcar sampling gate. (*a*) Circuit diagram, (*b*) signal form, (*c*) track switch operation, (*d*) memory function, (*e*) the memory function that would obtain if the switch were kept closed all the time.

Apart from the resistor R, the boxcar gate is identical to the track-and-hold gate of figure 8.9, and when the signal source resistance is allowed for, as in figure 8.10, the circuits become identical. However, in the boxcar gate, R is increased deliberately.

Figure 9.14(*c*) shows how the TRACK switch is operated, to coincide with the signal pulses, first assumed square for simplicity. Then (*e*) shows the memory function that would obtain if the switch were closed all the time. When the switch is open the capacitor voltage will not change. As far as the capacitor is concerned, time 'stands still' over these periods. Thus the effect is to spread the actual memory function as shown in (d), and so the gate output is roughly equivalent to the average of the CR/T_w pulses prior to the current time. When the signal pulses are not square the limits of sampling should be placed as in figure 9.5. As always, the gate output can either be connected directly to a chart recorder for display or digitised.

The boxcar gate might well be used in place of the simple CR filter in the PSD realisation of optimally weighted pulse sampling in figure 9.9. Sometimes the duty cycle T_f/T_w for the signal pulse train is very large. A real analogue multiplier will output some degree of internally generated drift and offset and the simple averaging afforded by the CR filter alone will exaggerate this relative to the required signal pulse. The boxcar, switched to include only the output pulse from the multiplier, will eliminate this problem.

For more complex signal shapes, such as the antisymmetric doublet in figure 9.10, the boxcar can be modified in the same way as the integrate-and-hold circuit shown.

9.6 OFFSET, DRIFT AND 1/f NOISE

We now consider how the above methods of pulse sampling need to be modified when offset, drift and $1/f$ noise are added to the white noise.

Weighting functions

The appropriate modification to the weighting function is developed in figures 9.15 and 9.16. Figure 9.15 shows the obvious method for correcting a signal pulse sample for offset. The value of the baseline is sampled at some time prior to the signal pulse and this value is subtracted from the signal sample v_p.

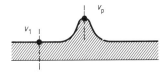

Figure **9.15** Offset correction for a sampled signal pulse (stored value $= v_p - v_1$).

When drift is present, reconstruction of the baseline requires two baseline samples as in figure 9.16(a). Here the baseline voltage v_{bp} for subtraction from v_p is given by

$$v_{bp} = v_1 \frac{(t_2 - t_p)}{(t_2 - t_1)} + v_2 \frac{(t_p - t_1)}{(t_2 - t_1)}. \tag{9.9}$$

Then, if t_1, t_2 and t_p are equidistant, $v_{bp} = (v_1 + v_2)/2$.

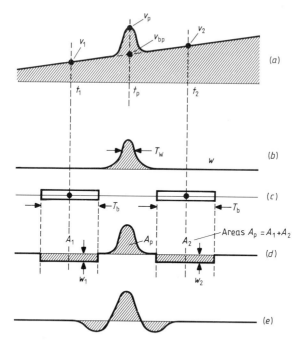

Figure 9.16 Modifications to the weighting function. (*a*) Drift correction, (*b*) weighting for white noise, (*c*) integration periods for baseline samples, (*d*) modified weighting function giving drift and offset correction, (*e*) optimum weighting for $1/f$ noise.

When white noise is present, point samples are not acceptable. Then the signal sample needs to be replaced by the weighted integral $\int wv \, dt$, with w as shown in (*b*). Also, the baseline samples will need to be replaced by integrals taken over a finite period T_b as in (*c*). This amounts to modifying the weighting in (*b*) to that in (*d*).

Since the white noise error in an average taken over a period T_{av} is proportional to $T_{av}^{-1/2}$, the noise error in the baseline samples will be prevented from adding significantly to the error in the signal pulse measurement if $T_b \gg T_W$ as shown.

Criteria for offset and drift rejection

In order to determine the precise weighting required for the baseline correction integrals in (*d*), it is necessary to establish the general criteria for a weighting function w to reject offset and drift.

Let the baseline voltage representing constant rate drift and offset be v_b. Then

$$v_b = A + Bt. \tag{9.10}$$

When the weighted integration is applied to v_b this gives

$$\int wv_b \, dt = A \int w \, dt + B \int wt \, dt. \tag{9.11}$$

This is zero if both

$$\int w \, dt = 0 \tag{9.12}$$

and

$$\int wt \, dt = 0. \tag{9.13}$$

Here equation (9.12) is the condition for offset rejection and (9.13) the condition for drift rejection.

Thus in figure 9.16(d) we require area $A_1 + A_2 = A_p$ for offset rejection. Condition (9.13) for drift rejection is that the first moment of w should be zero. For a symmetric signal peak this is most simply achieved if t_1, t_2 and t_p are equidistant and area $A_1 = A_2$. For an asymmetric pulse, either t_1 or t_2 may be adjusted or the balance between w_1 and w_2 in (d) altered.

Frequency response
When $1/f$ noise is present, a spectral view is required, and we need to determine the frequency response corresponding to the weighting function of figure 9.16(d). To this end we suppose the weighting to be realised by a filter and the filter to be divided as in figure 9.17(a). Filter (1) is the matched filter for white noise and so has a frequency response (c) that matches the signal pulse spectrum (b). Filter (2) is almost identical to a running average filter with averaging period $2T_b$. This gives the approximate frequency response in (d), with a cut-off frequency $f_b = (1/4\pi T_b)$. Responses (c) and (d) then combine to produce (e). Here the cancellation below the lower cut-off is evident because the integral of the weighting function w in figure 9.16(d) is zero, and so any component of period which is large compared with the overall width $2T_b$ will be attenuated.

It was shown above that, in order to avoid significant increase in noise error from the baseline correcting integrals, we require $T_b \gg T_w$. This is a time-domain view. The corresponding frequency-domain view now emerges from figure 9.17. If $T_b \gg T_w$ then the modified frequency response of (e) does not differ significantly from the white noise matched response of (c).

When $1/f$ noise is also present the position is altered. Without $1/f$ noise we require $T_b \gg T_w$ and there is no limit to how far T_b may exceed this condition. Normally one would make $T_b \simeq 10T_w$. However when $1/f$ noise is present, as in (f), it is necessary that the lower cut-off f_b should not fall below the noise corner frequency f_k. Where $f_k < f_w/10$ this involves no compromise. For $f_w/10 < f_k < f_w$ however, f_b must be increased to equal f_k and this does begin to compromise the white noise matching. Finally, if $f_k > f_w$ the signal spectrum is

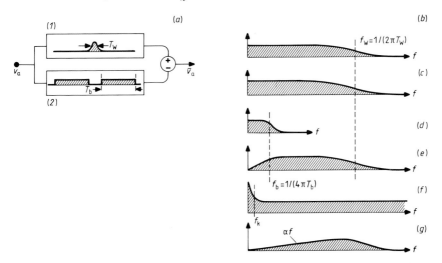

Figure 9.17 Frequency responses corresponding to the weighting functions of figure 9.16. (*a*) Filter, (*b*) signal spectrum, (*c*) frequency response to filter (*1*), (*d*) frequency response to filter (*2*), (*e*) frequency response of total filter, (*f*) noise spectrum at filter input, (*g*) frequency response for weighting function of figure 9.16(*e*), $1/f$ noise.

in the $1/f$ noise dominated part of the noise spectrum and the case becomes effectively one where the noise is entirely $1/f$. For all but the last case, effective realisation is possible using the simple *CR* filter circuit described below, by direct transversal filter realisation of the weighting function or by software. For the $1/f$ noise dominated case the present analysis is inadequate. Thus, following the brief discussion of *CR* filter realisation, we extend below the concept of white noise matched filtering given in §9.4 to the case of non-white noise.

CR filter realisation for $f_k \ll f_w$

Figure 9.18 shows the simple *CR* filter realisation of the frequency response of figure 9.17(*e*). Here, for $f_k < f_w/10$, f_2 and f_3 can be anywhere between f_k and $f_w/10$, while for $f_w/10 < f_k < f_w$, f_2 and f_3 should equal f_k.

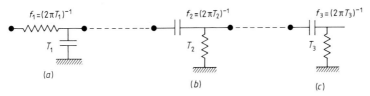

Figure 9.18 *CR* filter giving frequency responses of figure 9.17.

Two high-pass sections (b) and (c), rather than one, are required because, while it takes only one to reject offset, drift rejection requires two. For one section the steady-state condition is that the output voltage is constant. For an input drift rate dv_{in}/dt this drift then appears across the capacitor. The corresponding current $i = C\, dv_{in}/dt$, giving the constant output voltage $CR\, dv_{in}/dt$. Thus the drift is converted to offset and the second section is required to remove this offset.

The impulse response for the CR filter is developed in figure 9.19(a)–(c). It is interesting to note how this differs from that in figure 9.16(d). This is because the placings of t_1 and t_2 in figure 9.16 are arbitrary and in fact can be anywhere away from the signal. From the approximation of figure 9.19(d) it is clear that for the CR filter both t_1 and t_2 are on the same side of the signal peak. Then with the magnitudes and signs of the areas A_p, A_1 and A_2 as shown, it is still possible for the general conditions for drift and offset rejection to be observed. That is, for offset rejection the total area is zero and for drift rejection the first moment is zero.

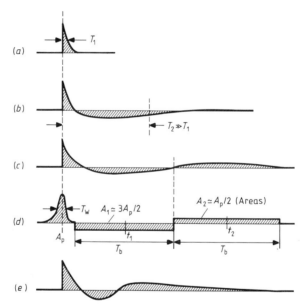

Figure 9.19 Impulse responses for CR filter of figure 9.18. (a) Filter (a). (b) Filter (a) + (b). (c) Filter (a) + (b) + (c). (d) Weighting function of figure 9.16(d) modified to bring t_1 and t_2 to the same side of the signal peak. (e) As (c), but with $T_2 = T_W$ to suit $1/f$ noise.

Matched filtering for non-white noise

The above arrangement is acceptable only as long as $f_W > f_k$. When this condition is violated and $1/f$ noise dominates we need to extend the concept of matched filtering to include non-white noise.

Figure 9.20 illustrates the general method. v_a in (a) is the noisy signal, composed of the non-white noise voltage v_n and the noiseless signal v_s. The noise spectrum chosen as an example is the realistic one of a mixture of white and $1/f$ noise, but the method is applicable to any noise spectrum.

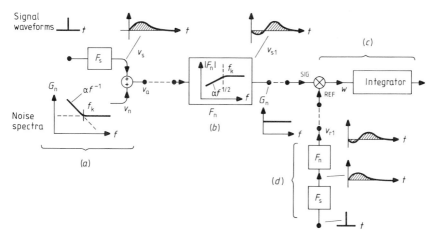

Figure 9.20 Processor for sampling a signal pulse in the presence of white noise. (a) Synthesis of noisy signal, (b) filter for making noise white, (c) analogue multiplier and gated integrator, (d) synthesis of weighting function v_{r1}.

The next step is to filter the noisy signal, using the filter F_n, to make the noise white. Since for $1/f$ noise the power spectral density $G_n \propto f^{-1}$, the correction needed will require the filter voltage gain $|F_n|$ to vary with $f^{1/2}$ in the $1/f$ noise dominated part of the spectrum, as shown in (b).

We now have a signal of known shape in the presence of white noise and the strategy for this has already been established. The signal must be weighted by a function w having the shape of the signal and then integrated. (c) shows the usual arrangement for this, following figure 9.6(b). Notice, however, that the 'reference' weighting signal v_{r1} should not have the shape of the original signal v_s but of v_{s1}, after the influence of F_n. Thus if v_{r1} is to be generated as the impulse response of a series of filters these will have to be F_s and F_n as for the signal arm. Hence the arrangement of (d).

An extension of the argument developed in §9.4 shows that it is possible to move any of the filter sections from one arm to the other, provided the phase-frequency response is reversed. If F_n is moved from the signal to the reference arm, the reference arm filter becomes $F_s F_n F_n^*$ as in figure 9.21. The weighting function is thus modified further and this is the function that would be used in the integrate-and-hold realisation of figure 9.7.

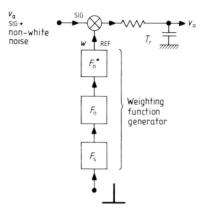

Figure 9.21 Transfer of F_n in figure 9.20 from signal arm to reference arm, to produce PSD-type realisation of figure 9.9.

It is equally possible to move all the filter sections from the reference to the signal arm as in figure 9.22, giving the modified transversal filter realisation. Here the transversal filter is identical to that in the reference arm of the PSD system of figure 9.21 apart from the phase reversal. Again the phase reversal is necessary to ensure the time reversal between the impulse response of the transversal filter and the weighting function produced by the reference arm filter.

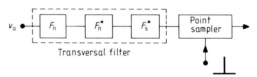

Figure 9.22 Transfer of F_s and F_n in reference arm of figure 9.20 to signal arm, to produce transversal filter realisation of figures 9.2 and 9.8.

Notice that for both realisations F_n is always combined with F_n^*. This combination gives zero phase shift at all frequencies and produces a combined voltage gain-frequency response $|F_n|^2$. With the combined $F_n F_n^*$ filter removed, the transversal filter system of figure 9.22 becomes the same as for white noise (compare figure 9.12). Notice however the way in which the two sections of $F_n F_n^*$ modify the noise spectrum, as shown in figure 9.23. The first section makes the noise white and then the second section applies this correction again. Also the voltage gain-frequency response becomes as in (b) with the gain $|F_n F_n^*| \propto f$ in the $1/f$ noise dominated part of the spectrum, rather than $\propto f^{1/2}$ as for F_n alone.

The spectral view of matched filtering, introduced in §9.4 for white noise alone, has now been extended. For white noise it was seen that the filter F_s^* had

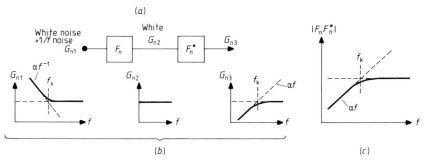

Figure 9.23 Diagram showing how the filter $F_n F_n^*$ in figure 9.22 reverses the noise-power spectral density G_n. (a) Filter, (b) G_n at various points in the filter, (c) voltage gain frequency response of filter $F_n F_n^*$.

two functions. The phase-frequency response was just that necessary to bring all of the spectral components of the signal into line at the point of sampling the filter output ($t = 0$). Also the gain-frequency response $|F_s^*|$ gave the spectral weighting necessary for minimum noise error.

Now for non-white noise the additional filter $F_n F_n^*$ is required. This has no phase shift and so does not upset the above phasing conditions. It does, however, modify the spectral weighting to suppress those components where the noise spectral density is highest. This is illustrated in figure 9.20 where the $1/f$ noise rising below the noise corner f_k in (a) is suppressed by the low-frequency cut-off in the frequency response of figure 9.23(c).

A precise expression for the gain-frequency response of the general noise correcting filter $F_n F_n^* F_s^*$ should now be given. The gain $|F_s^*|$ of filter F_s^* is proportional to the magnitude of the signal voltage components. Thus $|F_s^*| \propto G_s^{1/2}$, where G_s is the power spectral density of the signal. Also the voltage gain of the combination $F_n F_n^*$ is proportional to the noise power spectral density G_n. Thus, the overall voltage gain F of the noise correcting filter is given by

$$F \propto G_s^{1/2} G_n^{-1}. \tag{9.14}$$

In addition, the phase-frequency response of the overall noise-correcting filter will be the reverse of that for F_s, since $F_n F_n^*$ produces no phase shift.

Gaussian signal pulse
Figure 9.24 illustrates the above process for the case of a gaussian signal pulse and pure $1/f$ noise. This pulse shape is fairly representative and amenable to calculation. To obtain the weighting function, the gaussian signal pulse in (a) is first Fourier transformed to give its spectrum in (b). Then the $1/f$ noise noise-correcting factor f is applied in (c). (For the $1/f$ noise the spectral density $\propto f^{-1/2}$ so that the correction required is by $(f^{1/2})^2 \equiv f$.) Finally the corrected signal is re-transformed to give the weighting function shown in (d).

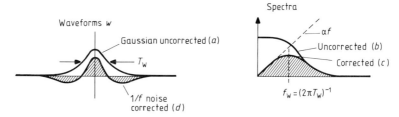

Figure 9.24 Derivation of optimum weighting w when sampling a gaussian pulse in the presence of $1/f$ noise alone. (*a*) and (*d*) waveforms, (*b*) and (*c*) spectra.

This is also shown in figure 9.16(*e*) and clearly corresponds to 9.16(*d*) with $T_b = T_w$. The corresponding modification to the spectrum of figure 9.17(*e*), taken from figure 9.24(*b*), is given in figure 9.17(*g*).

When white noise and $1/f$ noise are present, with $f_k \ll f_w$, figure 9.25 shows the appropriate modification to figure 9.24. The correspondence between the weighting functions of figures 9.25(*a*) and 9.16(*d*) and the frequency responses of 9.25(*b*) and 9.17(*e*) are evident.

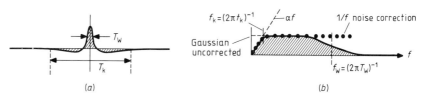

Figure 9.25 Modifications to weighting function of figure 9.24 for a mixture of white and $1/f$ noise. (*a*) Weighting function, (*b*) frequency response.

CR filter realisation for $f_k > f_w$

The frequency response of figures 9.17(*g*) and 9.24(*b*) for $1/f$ noise can be realised by the CR filter of figure 9.18 if the time constant of *one* of the high-pass sections is reduced to equal T_w. To reduce both would give a gain proportional to f^2, rather than the required f dependence. Figure 9.19(*e*) shows the corresponding impulse response, and it is clear that the correspondence with the required weighting of figure 9.24(*d*) is poor. This is largely because the added filter $F_n F_n^*$ producing the low-frequency cut-off should introduce no phase shift but the present CR filter introduces considerable shift. The correspondence is so poor that it is probably better to abandon attempts further to modify the CR filter, by adding extra sections, and instead use an arrangement such as in figures 9.26 and 9.27. Here only the low-pass section of figure 9.18(*a*) is needed and the overall memory function is constructed by taking three samples as shown and forming a suitably weighted sum. This weighting would most conveniently be done after digitisation.

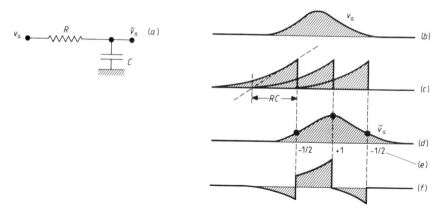

Figure 9.26 Approximate synthesis of $1/f$ noise corrected weighting of figure 9.24(b) using CR filter of figures 9.2 and 9.3, but with weighted three-point sampling. (a) Filter, (b) signal input, (c) memory function, (d) filter output, (e) sampling weights, (f) effective memory function.

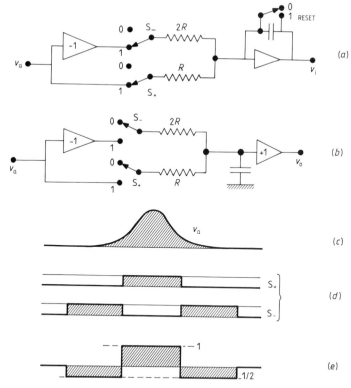

Figure 9.27 Approximate synthesis as in figure 9.26, but using (a) modified integrate-and-hold system of figure 9.4, (b) modified boxcar system of figure 9.14. (c) Input, (d) switching, (e) effective weighting.

Alternatively, where digitisation is not employed, the modified integrate-and-hold or boxcar systems of figure 9.27 will suffice, giving the effective weighting functions shown.

Doublet signal transient
When the signal transient is of the antisymmetric doublet form of figure 9.28(b) the position differs from the above in two respects. First, for the ordinary gaussian pulse of figure 9.24(a), if the frequency response is not modified as in (c) there is no attenuation of very low frequency $1/f$ noise components and the

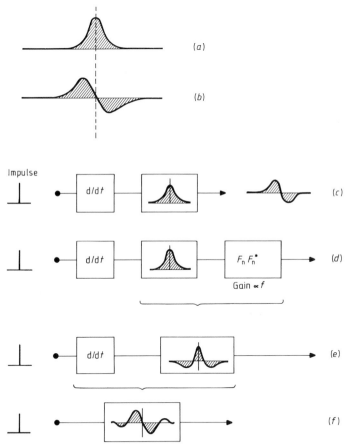

Figure 9.28 Development of weighting function for a gaussian doublet signal with $1/f$ noise. (a) Gaussian pulse, (b) doublet signal (time derivative of pulse (a)), (c) synthesis of weighting function for signal (b) in white noise, (d) addition of $1/f$ noise correcting filter, (e) combination of last two sections of (d), (f) combination of two sections in (e) to give required weighting function.

noise error is gross. In contrast, figures 9.29(*c*) and (*d*) show the optimum response for the doublet with and without the $1/f$ noise correction. Clearly neither accepts the low-frequency $1/f$ noise components, so the penalty for omitting the $1/f$ noise correction is modest.

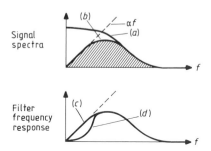

Figure 9.29 Diagram showing that, whether or not the matched filter for the gaussian signal doublet of figure 9.28(*b*) is corrected for $1/f$ noise, the low-frequency $1/f$ noise is still suppressed. (*a*) Gaussian pulse, (*b*) doublet, (*c*) uncorrected (proportional to f), (*d*) corrected for $1/f$ noise (proportional to f^2).

The second point is that the weighting giving white noise matching is the same as the shape of the signal. For the gaussian pulse, the first moment of this shape is zero, but for the doublet of figure 9.28(*b*) it is not. Thus for the doublet signal, with only the white noise matched filter, the system is exposed to drift.

Figure 9.28(*f*) shows the appropriately corrected weighting for $1/f$ noise, for which the first moment is zero. It is now clear that while this correction is modestly advantageous for $1/f$ noise it becomes more necessary when drift is prominent.

The details of figures 9.28 and 9.29 are developed as follows. Figure 9.29(*a*) shows the spectrum for the gaussian pulse. The doublet is the differential of this. A differentiator has a frequency response $\propto f$. Therefore the spectrum (*b*) for the doublet is obtained by scaling the gaussian spectrum (*a*) by f. (*c*) then shows the frequency response for white noise matching of the doublet, this being proportional to the doublet spectrum. For $1/f$ noise correction $|F_n F_n^*| \propto f$, so the $1/f$ noise matched response in (*d*) is given by scaling the white noise matched response (*c*) by a further factor f.

The $1/f$ noise corrected weighting function of figure 9.28(*f*) is developed in figure 9.28. The transition from (*d*) to (*e*) follows from figure 9.24.

It is clear that in (*f*) $\int w \, dt = 0$, so there is immunity to offset. That $\int wt \, dt = 0$ is less clear. However the differentiator in (*d*) converts any input drift to offset. The $1/f$ noise correction section then removes this offset because for this section the gain is zero at zero frequency.

Where drift is not prominent, the advantages of $1/f$ noise correction are barely worthwhile and the uncorrected systems of figures 9.10, 9.11, etc described in §9.3 or the modified boxcar of §9.5 constitute suitable realisation. When drift is prominent, and total drift rejection is necessary, the most simple realisation is to use a single CR filter as in figure 9.11 with $RC = T_W$ and to take four samples as in figure 9.30. With the weights chosen, both sum and first moments are zero. Similar extensions of the methods of figure 9.27 are also clearly possible. These constitute approximations to the more direct realisations of weighted integrate-and-hold and transversal filter, which remain as less economic possibilities.

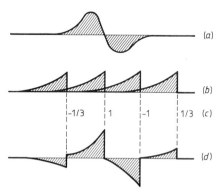

Figure 9.30 Use of the multi-point sampling technique of figure 9.26 with a doublet signal, rather than a signal pulse. (a) Signal input, (b) filter memory function, (c) sampling weights, (d) effective memory function, approximating to figure 9.28(f).

9.7 PEAK AND VALLEY DETECTORS

Sometimes it is wished to measure the maximum or minimum value that a signal attains over a given period. This is a form of sampling, in that the value required is that of the signal at one point in time.

Simple circuits

Figure 9.31 illustrates the operation of some suitable circuits. For the simple peak and valley detectors of (a) and (b), v_p and v_v represent the maximum and minimum values attained by the input voltage v_{in} at any time after the time t_1.

The waveforms in (c) assume the diodes to be ideal. In reality each diode requires a finite forward voltage v_d across its terminals in order to pass the charging current into the corresponding capacitor. Then v_p will be below and v_v above the true maximum and minimum values of v_{in} by the values of v_d for the two diodes. The precise value of v_d will depend upon dv_{in}/dt and therefore will be uncertain. Values somewhere between zero and approximately 600 mV may be expected.

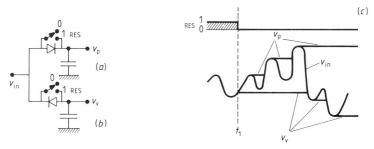

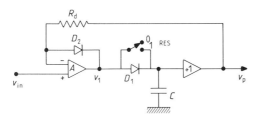

Figure 9.31 Simple peak and valley detector circuits. (*a*) Peak detector, (*b*) valley detector, (*c*) waveforms.

Feedback circuit

The feedback circuit of figure 9.32 allows the error due to v_d to be greatly reduced. Ignore R_d and D_2 for the moment, by assuming that $R_d = 0$ and that the diode is an open circuit. Suppose that v_p is initially zero and v_{in} is rising from zero. v_1 then becomes positive, causing D_1 to conduct, thereby completing the negative-feedback loop.

Figure 9.32 Feedback peak detector, which overcomes forward voltage drop of diode D_1.

For the operational amplifier

$$v_1 = A(v_{in} - v_p) \tag{9.15}$$

and for the diode D_1, the capacitor C and the isolator

$$v_1 = v_d + v_p. \tag{9.16}$$

Thus

$$v_p = (v_{in} - v_d/A)/(1 + A^{-1}). \tag{9.17}$$

Then, since $A \gg 1$, v_p follows v_{in} closely and the error due to v_d is reduced by the factor A^{-1} relative to that for the simple circuit.

When v_{in} stops rising and begins to decrease, the diode D_1 becomes reverse-biased, the capacitor C holds its charge and v_p remains stable at the value of the maximum. This continues until v_{in} rises to a value above v_p, at which stage v_p rises to assume the new maximum.

The purpose of R_d and D_2 is to stop the operational amplifier from saturating during the periods when v_p is 'holding'. Otherwise, v_1 then assumes the negative saturation value for the op-amp, causing the recovery time at the next point where v_{in} rises above v_p to be unnecessarily long.

To obtain the feedback valley detector, the diodes D_1 and D_2 are simply reversed.

Amplitude measurement for a pulse of unknown time of occurrence

As used in figure 9.33, the peak detector provides a means of recording the magnitude of a signal pulse when the time of occurrence of the pulse is unknown. This in contrast to all of the pulse sampling methods discussed so far, for which the time of occurrence of the pulse is known.

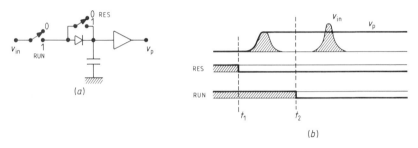

Figure 9.33 Use of peak detector to record the height of a pulse occurring at some unknown time between times t_1 and t_2. (a) Circuit diagram, (b) waveforms.

Provided that it is known that the pulse will occur between two times t_1 and t_2, the peak detector can be reset at t_1 and sampled at t_2. Alternatively, the run/hold switch can be operated as shown to hold the sample. The feedback circuit of figure 9.32 can be modified in the same way.

White noise

When white noise is present the signal must be weighted by its own shape and integrated before detection. Since the timing of the signal is unknown, the integrate-and-hold and boxcar systems are inapplicable and matched filtering must be used, as in figure 9.34. The diagram shows the filter output $\overline{v_a}$ as symmetric about the time t_s at which the filter memory function coincides with the signal, which is the required time of sampling. This point is vital and should be established.

The times t_1 and t_2 in (d) are equidistant from t_s. $\overline{v_a}$ is obtained by point-by-point multiplication (convolution) of memory function and input signal and will therefore be the same for t_1 and t_2. Thus $\overline{v_a}$ must be symmetric.

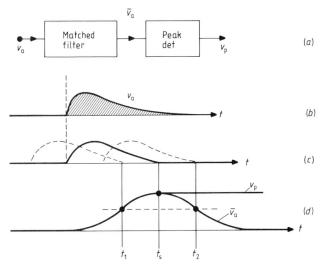

Figure 9.34 Use of matched filter when pulse in figure 9.33 is accompanied by white noise. (*a*) Block diagram, (*b*) input, (*c*) filter memory function, (*d*) output.

With the symmetry shown, $\bar{v}_a$ is maximum at the required time of sampling and so the peak detector records the sample taken at the correct time.

CR filter approximation

Once again, the matched filtering may be realised accurately by the appropriate transversal filter and approximated to by a simple low-pass *CR* filter with $RC = T_w$, the signal pulse width. This assumes a simple unidirectional pulse. For doublet and more complex shapes the simple *CR* filter will not suffice. Further, because of the unknown signal timing, the multiple sampling techniques of figures 9.26 and 9.27 cannot be applied. Thus the modified transversal filter seems to be the only approach possible.

Offset, drift and $1/f$ noise

When these are present the filter should be modified as in figure 9.16. The filter is now no longer strictly 'matched' as for white noise and our argument for the symmetry of $\bar{v}_a$ needs confirmation. From figure 9.20 it was shown that the effect of $F_n F_n^* F_s^*$, the modified matched filter, is to bring all the spectral components of the signal into phase at the time of sampling. Since these are all cosine components, the filter output will be symmetric. Thus the peak detector again records the sample at the correct time.

Limiting signal-to-noise ratio

Unfortunately the above arguments assume the application of the noiseless signal v_s to the matched filter. When noise is present this will displace the time of sampling by the peak detector. For noise levels such that the signal-to-noise ratio at the filter output is large compared with unity the displacement is small. Moreover it will be shown in §10.1 that the matched filter ensures that the timing displacement is the minimum possible. However, the effect clearly becomes catastrophic when the signal-to-noise ratio for $\bar{v}_a$ falls below unity. It therefore appears that this marks the threshold of our ability to measure the amplitude of a pulse of unknown timing in the presence of noise.

9.8 SOFTWARE REALISATION OF WEIGHTED PULSE SAMPLING

For any of the above methods, it is possible to obtain the weighted signal pulse integrals by logging the noisy signal into memory and forming the integral by software. There are two problems with this approach, to limit the amount of storage space required and to obtain sufficient speed of computation. These are to some degree linked. If sufficient memory space is available to store the entire noisy input signal then extended computation time is merely a matter of inconvenience. If, on the other hand, this amount of memory is not available, then the processing of each pulse must be done before the arrival of the next, and only the calculated pulse amplitude stored.

The main factor causing the computation time to be long is the extensive series of multiplications required in forming a weighted integral. This problem is much eased if the weighting function w has only the values 0, $+1$ and -1, as in the approximations of figure 9.35(b). The slightly more sophisticated approximation of (c) is possible with little extra increase in computation time if the values $\pm\frac{1}{2}$, $\pm\frac{1}{4}$, $\pm\frac{1}{8}$ etc are also allowed.

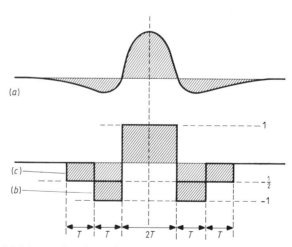

Figure 9.35 Approximation to ideal weighting function for rapid software processing. (a) Ideal weighting, (b) approximation allowing $w = +1$, 0, -1, (c) approximation allowing $w = +1$, $+\frac{1}{2}$, 0, $-\frac{1}{2}$, -1.

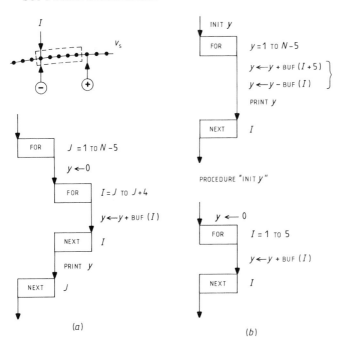

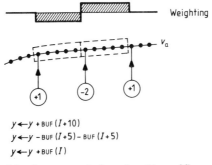

Figure 9.36 Algorithms for forming (and printing) a 5-point running average y of the signal v_a, where v_a is held in array BUF of length N. (a) Slow, (b) fast algorithm.

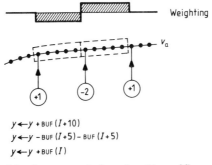

$$y \leftarrow y + \text{BUF} (I+10)$$
$$y \leftarrow y - \text{BUF} (I+5) - \text{BUF} (I+5)$$
$$y \leftarrow y + \text{BUF} (I)$$

Figure 9.37 Modification to step in fast algorithm of figure 9.36(b) to give doublet weighting.

When the signal timing is unknown, the potential computation time is vastly increased. This is because the weighted running integral corresponding to the weighted transversal filter is required. Software implementation of the integrate-and-hold circuit, in contrast, requires only one integration. In this situation, the simplest of approximations, such as the simple running average of figure 9.2, is all that will be possible. This is because only for the running average, or similar weightings where only values of 0, $+1$ and -1 are used, is the speed advantage of replacing the algorithm of figure 9.36(a) with that of (b) obtained. Figure 9.37 shows the extension of the method to a signal of more complex shape, the doublet.

The above methods are straightforward when the entire pulse train can be stored in memory. Where this is not possible a queue will suffice, where the length of the queue is enough to accommodate the number of samples within the average.

10 Measurement of the time of occurrence of a signal transient

In the last chapter, we considered how to measure the amplitude of a signal transient of known shape and usually known time of occurrence. In this chapter, we consider the associated problem of how to measure the time of occurrence of the transient, again assuming the shape is known.

First, in §10.1, the essentials of a general method are developed which allows the timing to be measured with the minimum error due to white noise. It is found that the approach differs slightly for the case of a 'pulse' where the voltage at either side of the transient is the same and a 'step' where it is not. Of the two, the pulse method is the more valuable, since it is found to be much less susceptible to offset, drift and $1/f$ noise. Thus, where possible, the signal to be timed should be made a pulse and not a step. For this reason, greatest emphasis is laid on the realisation of the pulse timing method. In §10.2 various realisations are covered, including the phase-locked loop, which is suited to a periodic train of such transients. It is found that, in making simple approximations to the ideal method, signal timing is subject to certain hazards that produce gross additional error. This is also discussed in the section, together with the necessary precautions needed to avoid the effect.

In §10.5 the modifications to the general method needed to overcome the effects of offset, drift etc are discussed. These are mostly applicable to the step because the pulse is not very subject to these effects anyway.

Section 10.6 describes the lock-in amplifier which combines the PSD and phase-locked loop methods to provide a system which is capable of both timing and measuring the amplitude of a periodic signal which is submerged in noise. Finally, §10.7 describes the autocorrelation method, which allows the optimum timing methods to be applied to a signal for which the detailed shape of the signal is initially unknown.

10.1 METHOD FOR TIMING A TRANSIENT WITH MINIMUM WHITE NOISE ERROR

Figure 10.1 shows the transient that is to be timed, with v_a representing the transient as seen at the output of the signal amplifier, with the noise superimposed, and v_s representing the 'unseen' signal component of v_a. Initially it is assumed that everything is known about v_s except its placing on the time axis. It is, therefore, possible to construct the 'reference' function v_r and to slide it along the time axis until the best 'fit' is obtained.

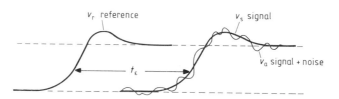

Figure 10.1 General signal transient to be timed with minimum white noise error.

Let t_ε be the time difference between v_r and v_s, as shown, and let $t_{\varepsilon f}$ be the value of t_ε that results when the curve-fitting has been carried out. Were there no noise, $t_{\varepsilon f}$ would be zero, but with the noise $t_{\varepsilon f}$ will be finite. The problem now is to find the curve-fitting criterion that will minimise $t_{\varepsilon f}$. $t_{\varepsilon f}$ is a random variable and it is therefore the standard deviation σ_ε for $t_{\varepsilon f}$ that must be found and minimised.

Figure 10.2 shows a simple technique for fitting the reference v_r and observed signal function v_a. Here a point on the reference function is selected arbitrarily and the reference function is slid along the time axis until the time $t_{ar} = 0$. A better fit can be expected if the time differences are determined for many points on the reference function, also as shown in figure 10.2. Then the fitting criterion is that $\sum_i t_{ai} = 0$. A reduced error is expected because the noise errors in the t_{ai} will average to a lower value than that expected for just one measurement.

The approach suffers from two disadvantages. First, it is inconvenient to measure a large number of time differences; generally voltages are easier to

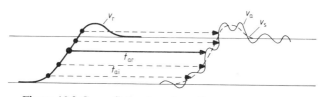

Figure 10.2 Curve fitting based on time measurements.

measure than times. Second, when the signal-to-noise ratio is poor it is quite possible that multiple crossings of the level corresponding to the chosen point on the reference function will give multiple values for each t_{ai}.

Where the signal transient v_s is a simple step, an obvious curve-fitting criterion is that

$$\int_{-\infty}^{+\infty} (v_a - v_r)\, dt = 0. \tag{10.1}$$

Figure 10.3 shows v_a and v_r fitted according to this criterion. The implication is simply that the area between the curves above and below v_r is equal. There is, further, no question of multiple values occurring for the variables under the integral.

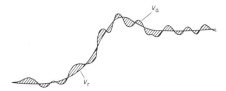

Figure 10.3 Curve fitting based on voltage measurements.

The criterion has one serious drawback when signals other than a simple step are involved. If v_n represents the noise superimposed on the signal transient v_s to constitute the observed voltage v_a, then

$$v_a = v_s + v_n. \tag{10.2}$$

Then the curve-fitting criterion of equation (10.1) can be written as

$$\int_{-\infty}^{+\infty} v_n\, dt = \int_{-\infty}^{+\infty} (v_r - v_s)\, dt. \tag{10.3}$$

Thus the time error t_{ef} arises as the reference function v_r is moved away from the signal function v_s enough for the integral $\int_{-\infty}^{+\infty} (v_r - v_s)\, dt$ to balance the integrated noise.

Figure 10.4 now shows the result of moving v_r away from v_s upon the integral. Notice that the integral before the peak of the curve is of the opposite sign to that after the peak. Thus, the reference curve has to be moved further when the integral after the peak is included than it would if that part of the integral had been totally ignored. In the extreme case of a signal pulse, rather than a step, the effect is total. Initially, sliding v_r away from v_s then has no effect at all on the integral $\int_{-\infty}^{+\infty} (v_r - v_s)\, dt$ and the resulting value of t_{ef} becomes very large.

It is clear that all that is required to correct the situation, and to make both halves of the integral work together, is to weight the value of $v_r - v_s$ with the sign of the slope of the reference function. As applied to the observed signal,

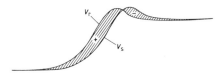

Figure 10.4 Diagram showing how $\int (v_r - v_s)\, dt$ is of opposite sign on either side of the signal peak.

this would mean the 'fitting' criterion

$$\int_{-\infty}^{+\infty} (v_a - v_r)S\, dt = 0 \tag{10.4}$$

where $S = \pm 1$ according to the sign of dv_r/dt.

The question of weighting the values of $v_a - v_r$ can now be taken further. Suppose initially that, instead of using the integral of equation (10.1) as the curve-fitting criterion, fitting was based on the comparison of v_a and v_r at one point on the reference function only. The reference is then moved along the time axis until

$$v_a - v_r = 0 \tag{10.5}$$

and the result is shown in figure 10.5.

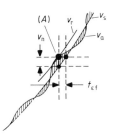

Figure 10.5 Diagram showing that when fitting is at one point (A) only, the timing error $t_{\varepsilon f} = v_n (dv_r/dt)^{-1}$.

From the diagram, the timing error $t_{\varepsilon f}$ is given by

$$t_{\varepsilon f} = v_n (dv_r/dt)^{-1}. \tag{10.6}$$

Then, if $\tilde{v}_n$ is the RMS value of the noise voltage v_n, the standard deviation σ_ε for $t_{\varepsilon f}$ will be given by

$$\sigma_\varepsilon = \tilde{v}_n (dv_r/dt)^{-1}. \tag{10.7}$$

To minimise σ_ε it is then seen to be advantageous to place the point on the reference function v_r where comparison is made with the signal function v_a at the point of maximum slope. It is also clear that dv_s/dt, the signal slope, which is equal to dv_r/dt, should be made as large as possible. In fact, a large slope for the signal is more important than for the signal itself to be large.

It now seems that, when the single-point comparison of equation (10.1) is extended to the multiple-point comparison of equation (10.5), it would be advantageous to weight the values of $v_a - v_r$ in favour of those points where the slope dv_r/dt of the reference function is high. A weighting in direct proportion to dv_r/dt would give the curve-fitting criterion

$$\int_{-\infty}^{+\infty} (v_a - v_r)(dv_r/dt)\,dt = 0. \tag{10.8}$$

Note that this weighting also incorporates the sign of dv_r/dt, as seen to be necessary for other than a signal step.

It is now shown that the weighting of equation (10.8) is, in fact, the best one to use. We start with a general weighting function w, giving the criterion

$$\int_{-\infty}^{+\infty} (v_a - v_r)w(t)\,dt = 0 \tag{10.9}$$

and show that $w = (dv_r/dt)_t$ is the weighting function that gives the minimum value for σ_ε, the standard deviation for the time error.

From equations (10.2) and (10.9)

$$\int_{-\infty}^{+\infty} v_n w\,dt = \int_{-\infty}^{+\infty} (v_r - v_s)w\,dt. \tag{10.10}$$

But the RHS $= t_{\mathrm{ef}} \int_{-\infty}^{+\infty} (dv_r/dt)_t w\,dt$ so that

$$t_{\mathrm{ef}} = \frac{\int_{-\infty}^{+\infty} v_n w\,dt}{\int_{-\infty}^{+\infty} (dv_r/dt)_t w\,dt}. \tag{10.11}$$

Now the noise function v_n shown in the diagrams so far has assumed some sort of low-pass filtering. Otherwise, for white noise v_n would be infinite. The assumption was made to aid visualisation but is not necessary for the present argument and, in fact, complicates it. Thus, in our attempts to evaluate and minimise σ_ε, we assume the noise to be initially unfiltered. To overcome the problem of infinite v_n, the time axis is divided into equal intervals of length Δt, where Δt is sufficiently small for the weighting function w to be effectively constant over the interval. The numerator of the expression for t_{ef} can then be expressed as $\sum_i v_{n\,\Delta i} w_i\,\Delta t$, where $v_{n\,\Delta i}$ is the average of the unfiltered noise v_n taken over the ith interval of width Δt. Also, w_i is the value of w for that interval. Unlike v_n, $v_{n\,\Delta i}$ is finite. In a similar fashion, it is convenient to express the denominator as $\sum_i (dv_r/dt)_i w_i\,\Delta t$, giving

$$t_{\mathrm{ef}} = \frac{\sum_i v_{n\,\Delta i} w_i}{\sum_i (dv_r/dt)_i w_i}. \tag{10.12}$$

Now let $\sigma_{\varepsilon i}$ represent the value σ_ε when only the noise over the ith interval is present. Then from equation (10.12)

$$\sigma_{\varepsilon i} = \frac{\sigma_{n\,\Delta i} w_i}{\sum_i (dv_r/dt)_i w_i} \tag{10.13}$$

where $\sigma_{n\,\Delta i}$ is the standard deviation for $v_{n\,\Delta i}$.

Since the time errors associated with the noise over each interval are unrelated

$$\sigma_\varepsilon^2 = \sum_i \sigma_{\varepsilon i}^2. \tag{10.14}$$

Note also that all the $\sigma_{n\,\Delta i}$ are equal, since the intervals are all of equal length. Let $\sigma_{n\,\Delta i} = \sigma_{n\Delta}$ for all i.

Then, combining equations (10.13) and (10.14) gives

$$\sigma_\varepsilon = \frac{\sigma_{n\Delta}(\sum_i w_i^2)^{1/2}}{\sum_i (dv_r/dt)_i w_i}. \tag{10.15}$$

An expression for σ_ε has now been obtained, and it is now required to find the values of the weights w_i that make σ_ε minimum. We select one weight, w_j, and find the condition for which

$$\frac{\partial \sigma_\varepsilon}{\partial w_j} = 0. \tag{10.16}$$

The resulting condition is that

$$w_j = \frac{\sum_i w_i^2}{\sum_i w_i (dv_r/dt)_i} (dv_r/dt)_j \tag{10.17}$$

and this is satisfied if, for all i, including $i = j$, $w_i = (dv_r/dt)_i$. Thus the weighting function w in equation (10.9) should equal dv_r/dt and this confirms equation (10.8) as being the best curve-fitting criterion.

10.2 SIGNAL PULSE TIMING

The next step is to apply the criterion of equation (10.8) to the timing of a signal pulse. The equation can be written in the form

$$\int_{-\infty}^{+\infty} v_a(dv_r/dt)\,dt = \int_{-\infty}^{+\infty} v_r(dv_r/dt)\,dt. \tag{10.18}$$

But the RHS of the equation is equal to $[v_r^2/2]_{t=-\infty}^{t=+\infty}$ so the criterion becomes

$$\int_{-\infty}^{+\infty} v_a(dv_r/dt)\,dt = [v_r^2/2]_{t=-\infty}^{t=+\infty}. \tag{10.19}$$

For the pulse, v_r for $t = -\infty$ and $t = +\infty$ are the same, so the curve-fitting criterion becomes

$$\int_{-\infty}^{+\infty} v_a(dv_r/dt)\, dt = 0. \qquad (10.20)$$

Figure 10.6 now shows the pencil and paper interpretation of applying the criterion to the signal pulse. Here (a) shows the reference function v_r placed at some arbitrary position on the time axis, separated from the unseen signal function v_s by the time difference t_ε. Figure 10.6(b) then shows the required weighting function dv_r/dt corresponding to v_r. The integral $I = \int_{-\infty}^{+\infty} v_a$ $(dv_r/dt)dt$ corresponding to the LHS of the curve-fitting criterion of equation (10.20) can then be formed, graphically, if necessary, using the diagrams (a) and (b). It is then possible to slide v_r and the associated dv_r/dt function along the time axis causing I to vary as in (c). Then the time at which I in (c) crosses zero is the time at which the reference function v_r in (a) is fitted to the noisy signal pulse v_a according to the required criterion and therefore with the minimum white noise error.

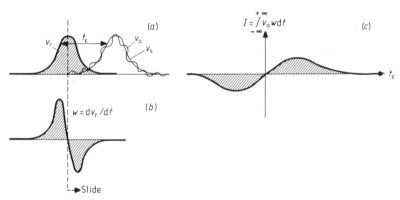

Figure 10.6 Pencil and paper representation of optimum pulse timing. (c) Variation of I with t_ε as v_r (a) and w (b) are slid along the time axis.

Transversal filter realisation

There are a variety of ways of realising the above 'pencil and paper' process, either by hardware or software. We shall consider two analogue circuit methods and the first is the transversal filter method shown in figure 10.7(a)–(e). Here the weighting function w in figure 10.6(b) is provided by the transversal filter in figure 10.7(a). The filter output v_o then represents the integral I in figure 10.6(c). Then the zero crossing in v_o represents the time at which the filter function w in (c) matches the signal in (b), as shown. The zero crossing in v_o is identified by the comparator, which produces a corresponding transition in the logic output c. This can then be timed by a digital timer or whatever other method is convenient. The delay $T_1/2$ is inevitable and can be allowed for, since T_1 is known.

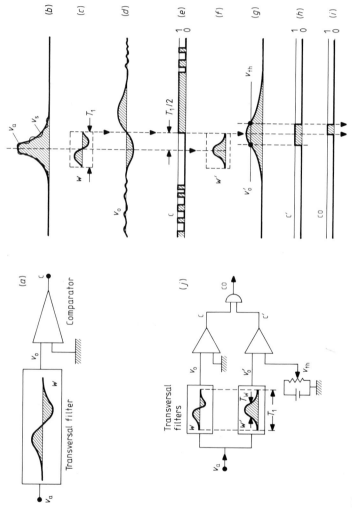

Figure 10.7 Transversal filter realisation of the pulse timing process of figure 10.6. (*a*) Basic system, (*b*)–(*i*), see text, (*j*) modification avoiding false zero crossings.

A minor complication with the arrangement is that, away from the signal pulse, the noise produces series of 'false' zero crossings in v_o. These may be seen in the wings of figure 10.7(d) and the corresponding effect upon the comparator output c in (e). The solution shown is to use a second transversal filter as in (j) in order to determine the amplitude of the signal, again with the minimum noise error. The filter output v_o' is then compared with the threshold level v_{th} and the logic ensures, in the way illustrated in the waveforms (f)–(i), that the main output co only registers zero crossings of v_o when v_o is above the threshold.

Transient generator realisation

As for pulse amplitude measurement, the pulse timing measurement can be realised using a transient generator and analogue multiplier as in figure 10.8. Here, the weighting function w in figure 10.6(b) is provided by the transient generator and the final integration by the low-pass filter. Then, when the weighting function is matched in time to the signal pulse, as shown, the integral of the product v_p is zero.

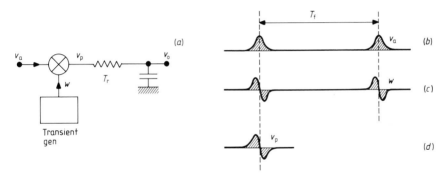

Figure 10.8 Transient generator realisation of pulse timing process of figure 10.6.

Unlike the above transversal filter method, the present method is only suitable for a periodic pulse train. The time constant T_r of the low-pass filter following the multiplier is made large compared with the pulse repetition period T_f. Then the low pass filter represents the average of v_p over the period T_r. Then, if the phasing of the w waveform is slowly changed relative to the signal waveform, v_o describes the function I in figure 10.6(c). Thus the procedure is to adjust the phasing until v_o is zero and then to take the timing from the oscillator that provides the w waveform. In this instance there is no time delay $T_1/2$ to allow for.

Phase-locked loop

The usual way of making this adjustment is shown in figure 10.9. Here the function w is generated by a voltage-controlled oscillator (vco). This is an oscillator for which the frequency f_0 is proportional to the control voltage v_c. In the absence of the input v_o, f_0 is set, using the control potentiometer, to a value as close to the signal frequency $f_f = T_f^{-1}$ as possible. Then, when v_o is connected, provided the sign is such as to make the feedback negative, the loop locks to a phasing close to that shown in figure 10.8.

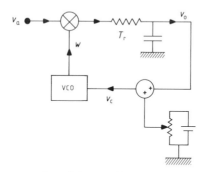

Figure 10.9 Incorporation of the transient generator system of figure 10.8 into a phase-locked loop.

The actual dynamics of the locking process are a little complex but it will be clear that if the phasing of w with respect to the signal starts to drift, the voltage v_o then applied to the vco control will alter f_0 for just long enough to correct the phase error. Actually the vco constitutes an integrator, in that the phase error is the integral of the frequency error. This means that in the final steady-state condition (forgetting noise for the moment) the frequency error must become zero. Thus the frequency lock is exact. The only phase error will be that needed to correct the small error in the setting of the control potentiometer and this can be made small enough to be insignificant.

More important than the locking dynamics for present purposes is that the system maintains the reference w in the position defined by the optimum timing criterion. Thus timing taken from the waveform w, by whatever means is convenient, probably digital counting, will be subject to the least noise error possible.

The process can also be realised using a transversal filter followed by a point sampling gate, much as for single transients. The filter and gate will replace the analogue multiplier in figure 10.9 and will have the advantage that the vco output needs only to be a train of point sampling impulses, rather than the profiled waveform w. The low-pass filter will still be needed in order to reduce the noise by averaging a series of the individual weighted pulse averages.

As usual, substitution of convenient approximations to the ideal weighting function can be made with very little increase in noise error. For example, the ideal doublet in figure 10.10(b) for a signal pulse can be approximated by the box approximation of (c). This can be done using the modified boxcar scheme in (a).

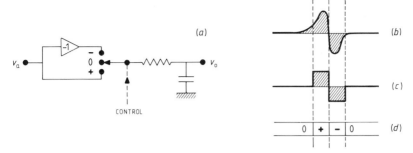

Figure 10.10 Use of modified boxcar circuit to provide a working approximation to the optimum weighting function for signal pulse timing. (*a*) System, (*b*) optimum weighting, (*c*) approximation to (*b*), (*d*) switch control.

Peak timing

Sometimes a signal pulse is timed by locating the time of occurrence of the maximum value. From our previous discussion it seems that this is a bad arrangement, because the slope is zero there and noise error in timing is inversely proportional to signal slope. However, the situation is altered when the signal is suitably filtered.

Figure 10.11(*a*) shows one way of synthesising the filter of figure 10.7(*a*) used with the usual zero crossing method. The memory function w of the zero crossing filter is the differential of the signal pulse shape, so the memory function w can be produced by a filter having the memory function w' which matches the signal pulse, and by following this by a differentiator. This is the arrangement shown in figure 10.11(*a*).

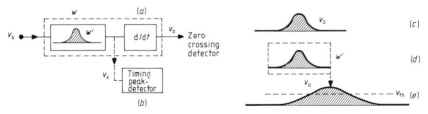

Figure 10.11 Diagram showing how zero crossing method of figure 10.7 can be replaced by a peak detector.

But the time at which v_o in figure 10.11(*a*) crosses zero is also the time at which v_x is maximum. Thus, providing the matched filter w is used, peak detection of v_x gives the same optimised timing error for white noise as does the zero crossing method.

It is doubtful which of the two systems is ultimately simpler. The peak timing system still requires the threshold v_{th} to avoid registering the noise peaks away from the signal pulse. It is possible that peak timing is a little easier to implement in software than is the zero crossing method but for a hardware implementation the zero crossing method is usually the simpler.

A suitable software routine for peak detection would be

$T=0$, $T_p=0$, $V_p=0$.
For each value of v_x input
   If $v_x > V_p$ and $v_x > v_{th}$
     Then $V_p=v_x$, $T_p=T$
  $T=T+1$.

Here T is a timer, T_p the time at which the peak occurred and V_p the peak value.

Hazard in peak detection

There is one none-too-obvious hazard associated with peak detection and this is illustrated in figure 10.12. Here the matched filter w' in figure 10.11 is replaced by the approximation of a simple single-section low-pass filter. This gives the memory function shown in figure 10.12(c), which is a reasonable approximation to the ideal, provided $T_r = T_w$. However, (d) shows the function w when the filter is followed by a differentiator for zero crossing detection as in figure 10.11(a). Now, this function includes an impulse, and this implies that the zero detected is the difference between the corresponding point sample of the noisy signal pulse v_a and the preceding average over T_r. But, for white noise, the error in a true point sample is infinite and so infinite error in the timing results. Since the time of occurrence of the peak is the same as the time of the zero crossing, the peak time too will be subject to the infinite timing error.

The prediction of infinite timing error is pessimistic, being the consequence of certain linear approximations made in the establishment of the original timing process of figure 10.6. However, it is clear that, unless some modification is made, the error will be grossly in excess of that for the ideal filter function.

The same problem occurs when a simple running average is used prior to peak detection timing. Figure 10.13(a) shows such a filter used to process the signal pulse in (d). Once again, the corresponding filter required for the zero crossing method would be that for peak detection, followed by a differentiator, as shown in (b). The memory function for the zero crossing filter then becomes as in (c). But this is equivalent to taking two point samples of the signal pulse and noting the time that they are equal.

In the absence of noise this is a perfectly acceptable way of finding the time of occurrence of the pulse but, again, when white noise is present the noise error becomes gross.

Clearly the solution is to expand the point samples into averages to give the memory function shown in (e). Here, to give the best approximation to the ideal function of figure 10.12(g), T_1 needs to be somewhat less than T_w, where

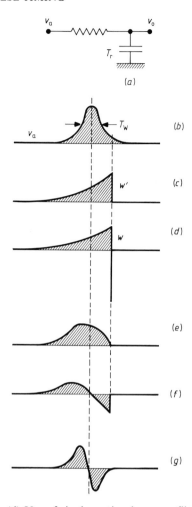

Figure 10.12 (a)–(d) Use of single-section low-pass filter in signal pulse timing by peak detection and zero crossing detection. (a) Filter, (b) noisy signal pulse, (c) memory function for filter, (d) as (c), but followed by a differentiator (as in figure 10.11(a) for zero crossing detection). (e)–(g) Use of two-section low-pass filter: (e) memory function, (f) memory function for two-section low-pass filter followed by a differentiator, (g) optimum weighting function for zero crossing detection.

T_W is the signal pulse width and also the period for the original running average.

For peak detection, the required memory function will be the integral of (e), which is shown in (f). Since (e) is obtained by forming the running average of (c) over T_1, it follows that (e) is obtained by following the filter in (b) by a further running average filter, and that (f) is obtained by removing the

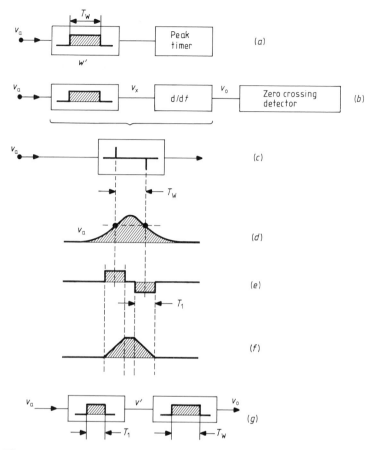

Figure 10.13 Use of running average filters in signal pulse timing. (*a*) Single running average with peak detection, (*b*) equivalent filter for zero crossing method, (*c*) memory function for filter in (*b*), (*d*) noisy signal pulse, (*e*) modified function in (*c*), (*f*) integral of function in (*e*), (*g*) use of cascaded running average filters to give the memory function in (*f*).

differentiator. Thus (*f*) is obtained by two cascaded running averaging filters as in (*g*).

Software implementation

Running average realisations of the above kind are more likely to be implemented in software than hardware. The usual way to maintain a running average v_0 over, say, the period T_1 is to apply the step

$$v_0 \leftarrow v_0 + v_a(t) - v_a(t - T_1)$$

in which $v_a(t)$ is the current sample of the input signal and $v_a(t - T_1)$ is the value

T_1 seconds earlier. This actually maintains the running sum rather than the strict running average of v_a, but the two are proportional, so the sum can be used just as well.

If two running averaging filters are to be cascaded, as in figure 10.13(g), one method would be to log all the samples into a buffer, to perform the first running average, possibly storing the values back in the same buffer, and then repeating the process for the second average. This is expensive in storage space and often 'real time' processing is required, where the filtered signal is produced as the logging takes place. Figure 10.14 illustrates a method where the only storage required is that necessary to retain samples over the period $T_W + T_1$ prior to the present.

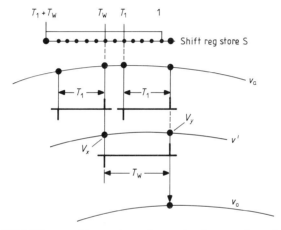

Figure 10.14 Diagram illustrating software implementation of the two-section running average filter in figure 10.13(g).

Once again, it is convenient to determine running sums rather than running averages. Thus, in figure 10.13(g) the final output v_o is the running sum of the first running sum v'. Thus to maintain v_o we need the present value $v'(t)$ and the earlier value $v'(t - T_W)$ of v'. Then to maintain v_o the step

$$v_o \leftarrow v_o + v'(t) - v'(t - T_W)$$

is required. For this $v'(t)$ and $v'(t - T_W)$ must also be maintained. This requires the steps

$$v'(t) \leftarrow v'(t) + v_a(t) - v_a(t - T_1)$$

$$v'(t - T_W) \leftarrow v'(t - T_W) + v_a(t - T_W) - v_a(t - T_W - T_1).$$

For this it is only necessary to store the values of the input signal v_a for the period $T_W + T_1$ prior to the present and this is most easily done in a software shift register (a 'queue'). The entire process is illustrated in figure 10.14, in which the variables $v'(t)$ and $v'(t - T_W)$ are represented by V_y and V_x respectively.

For the peak detection method, the time at which v_o is maximum is simply noted. For the zero crossing method the output required is the differential of v_o. But this is $V_y - V_x$ and so v_o does not actually have to be maintained. All that is required is to note when $V_x = V_y$.

CR filter realisation

It will by now be fairly obvious what is needed in order to overcome the above hazard in peak detection signal pulse timing when *CR* filtering is used. If the single-section *CR* filter in figure 10.12(a) is replaced by two such sections, the memory function becomes that in (e). When the differentiator is added, for the zero crossing method, the impulse in (d) is removed as in (f) and we are no longer taking a point sample of the input signal. Instead the overall memory function corresponds to the difference between two time averages of the input signal, each of period comparable with the pulse width T_W. Thus the gross additional error associated with taking a point sample of white noise is avoided. This applies equally to the peak as to the zero crossing methods because the final time given for both is the same. Also, with the filter response time T_1 equal to the signal pulse width T_W, a reasonably good approximation to the optimum weighting function for the zero crossing method in (g) is obtained.

Figure 10.15(a) shows the two-section low-pass filter, followed by the differentiator required when the zero crossing method is to be used. The frequency response curves in (b)–(d) give further insight into the reason for requiring a two-section, rather than a single-section filter. The single-section filter plus differentiator in (c) gives an overall frequency response that extends upwards to $f = \infty$. This admits an infinite amount of white noise and so gives gross timing error.

Another way of looking at the need for the two filter sections is to consider the signal pulse spectrum. This is shown in figure 10.16(a) for the signal pulse. The correct filter for the zero crossing method will have a memory function which is the differential of the pulse shape, and this will mean that the filter gain-frequency response will be proportional to the signal spectrum multiplied by f. The result, shown in (b), is clearly much better approximated by the two-section filter, for which the frequency response is shown in figure 10.15(d), than by the single-section filter, with the response in (c). Clearly, (c) is not even an approximate match.

A very simple way to add the differentiator to the two-section filter in figure 10.15(a) is to replace one of the low-pass sections by the equivalent high-pass section, just by reversing the C and R. Alternatively, the modified differentiator circuit of (e) gives the required overall response, for the zero crossing method.

It is interesting to consider the effect of reducing the filter time constant T_1 in figure 10.15 to a value much less than the required one of T_W, the signal pulse

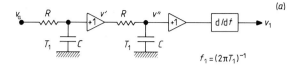

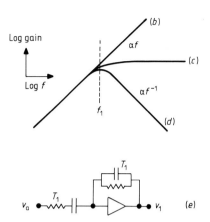

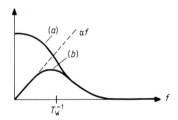

Figure 10.15 Two-section low-pass filter and differentiator giving memory function of figure 10.12(f). (a) Circuit diagram, (b)–(d) frequency responses, differentiator alone, with one and two low-pass sections respectively, (e) modified analogue differentiator giving frequency response of (a) and (d).

Figure 10.16 (a) Spectrum for gaussian signal pulse, (b) frequency response of filter for zero crossing signal timing.

width. Figure 10.17(b) shows the corresponding memory function. Then as $T_1 \to 0$ the process reduces to direct differentiation. This also is confirmed by the frequency response in figure 10.15(d) as $f_1 \to \infty$.

Thus, as $T_1 \to 0$ the process reduces to locating the peak of the unfiltered signal and this we now know is the worst possible place to determine the signal timing, because there the signal slope is least. However, by incorporating the CR filter and suitably adjusting T_1, the arrangement is converted from being fundamentally unsound to being the best possible. This amounts to identifying

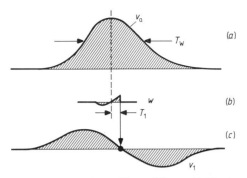

Figure 10.17 Operation of CR filter of figure 10.15 when $T_1 \ll T_W$.

the time when a pair of averages of the signal pulse of width T_W are equal, rather than identifying the time when the slope of the pulse is zero.

It will now be clear that, when approximating to the ideal realisation of the signal timing method using either the process of forming a running average or the simple CR filter, in order to avoid the hazard of effectively point sampling the signal it is necessary to use two stages rather than one. If CR filtering is used then two sections are required, while, if running averaging is used, the running average must be taken of the running average.

10.3 SIGNAL STEP TIMING

In this section, we consider how to time a signal step with minimum white noise error, rather than the signal pulse of the last section. Figure 10.18(a) shows the simplest type of signal step. However, it is in the interpretation of the optimised fitting criterion of equation (10.19) that the true distinction between a step and a pulse is given, at least for the present purposes. A pulse is defined as being any transient for which the value $[v_r^2/2]_{t=-\infty}^{t=+\infty}$ of the RHS of the equation is zero. Since v_r is essentially the same as the input signal v_s, this means a signal is a 'pulse' if its initial and final values well away from the transient are the same. For a 'step', on the other hand, the values are different. Thus a pulse can be superimposed upon any level of offset. It can also be of more complex shape than the simple gaussian function so far considered. All that is necessary is that the weighting function should equal the differential of the 'pulse' shape.

Similarly, the step now to be considered may be of more complex shape than the simple step in figure 10.18(a). One example is the generalised signal transient of figure 10.1.

With these reservations, we return to the simple step of figure 10.18(a). The object is to match the reference function v_r in (a) to the noisy signal v_a. Here, the correct matching criterion is to establish the weighting function w in (b), which is equal to dv_r/dt. Then the weighted integral $I_s = \int_{-\infty}^{+\infty} v_a w \, dt$ is formed and

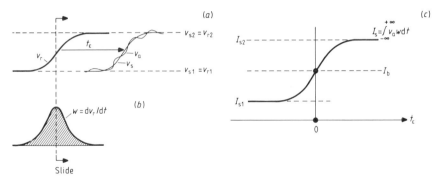

Figure 10.18 Pencil and paper representation of optimum step timing. (c) shows the variation of I_s with t_ε as v_r (a) and w (b) are slid along the time axis.

fitting occurs when this, the LHS of equation (10.19), is equal to the RHS $I_b = [v_r^2/2]_{t=-\infty}^{t=+\infty}$.

For the signal pulse, the RHS is zero, so the fitting is marked by the zero crossing of I_s. Here, however, the RHS is not zero and so the fitting is marked by the time when I_s crosses I_b.

Figure 10.18(c) shows the LHS I_s plotted against the time difference t_ε between the signal and reference functions in (a). It is a fairly simple matter to show that I_b is half-way between the initial and final values of I_s as t_ε is varied through zero. This assumes no noise is present.

For the initial value I_{s1} of I_s, and with no noise, $v_a = v_{s1}$, where v_{s1} is the initial value of the noise-free step, shown in (a). Thus

$$I_{s1} = \int_{t=-\infty}^{t=+\infty} v_{s1}\, dv_r = v_{s1}(v_{r2} - v_{r1}). \qquad (10.21)$$

Also for I_{s2}, $v_a = v_{s2}$ so

$$I_{s2} = \int_{t=-\infty}^{t=+\infty} v_{s2}\, dv_r = v_{s2}(v_{r2} - v_{r1}). \qquad (10.22)$$

But

$$I_b = \tfrac{1}{2}[v_r^2]_{t=-\infty}^{t=+\infty} = (v_{r2}^2 - v_{r1}^2)/2. \qquad (10.23)$$

Thus, since $v_{r1} = v_{s1}$ and $v_{r2} = v_{s2}$

$$I_b = (I_{s1} + I_{s2})/2. \qquad (10.24)$$

Figure 10.19(a) shows an appropriate method of hardware realisation. This is similar to the arrangement of figure 10.6(a) for the signal pulse, except that the filter now has the memory function proportional to the slope of the signal step, instead of the slope of the signal pulse. Also, the transition in the comparator output marks the time where the filter output is equal to v_b, rather than to zero. Here, to be compatible with I_b in figure 10.18(c), v_b is the mean of

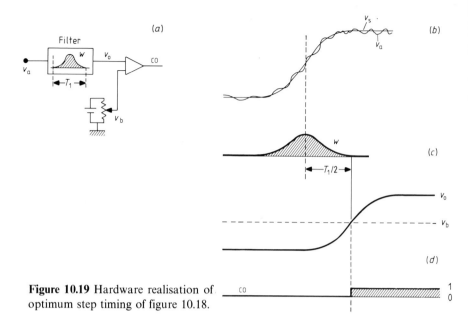

Figure 10.19 Hardware realisation of optimum step timing of figure 10.18.

the initial and final values of the filter output v_0. The arrangement again produces a time delay, here $T_1/2$, but again this is predictable and can be allowed for.

It is possible, as usual, to approximate the gaussian weighting function in figure 10.19(a) using a simple running average or, more probably, a simple CR low-pass filter of time constant equal to the step rise-time.

10.4 COMPARISON OF THE STEP AND PULSE METHODS

Having arrived at the optimum systems for timing both the step and the signal pulse, it is now desirable to calculate the actual white noise error for both types of system, so that they can be compared.

Equation (10.15) gives the standard deviation σ_ε for the timing error in either case. Inserting the optimised weighting function $w_i = (dv_r/dt)_i$ then gives

$$\sigma_\varepsilon = \sigma_{n\Delta} \bigg/ \left(\sum_i (dv_r/dt)_i^2 \right)^{1/2}. \tag{10.25}$$

A more convenient form would be

$$\sigma_\varepsilon = \frac{\sigma_{n\Delta}(\Delta t)^{1/2}}{\left[\int_{-\infty}^{+\infty} (dv_r/dt)^2 \, dt \right]^{1/2}} \tag{10.26}$$

where Δt is the time interval between each of the points in the summation of equation (10.25).

Note that the value of the numerator of the expression is independent of Δt, since $\sigma_{n\Delta}$, the standard deviation for the noise voltage averaged over the interval Δt, is proportional to $(\Delta t)^{-1/2}$.

For the idealised signal step of figure 10.20(a), equation (10.26) gives that

$$\sigma_\varepsilon = \sigma_{n\Delta}(\Delta t)^{1/2} T_r^{1/2}/V_s. \qquad (10.27)$$

The corresponding calculation for the pulse of figure 10.20(b) gives

$$\sigma_\varepsilon = \sigma_{n\Delta}(\Delta t)^{1/2} T_W^{1/2}/(2^{1/2} V_s) \qquad (10.28)$$

which, with $T_r = T_W$, gives almost identical noise errors.

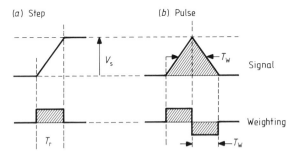

Figure 10.20 Simplified step and pulse waveforms for white noise error evaluation.

This, however, assumes white noise only. When offset, drift etc are considered the pulse timing has several advantages over the step. These are:

(1) The weighting function w_1 shown in figure 10.7(a) for the signal pulse has an integral of zero, while for the corresponding function w in figure 10.19(a) the integral is not zero. Thus, for the step, the system is susceptible to offset, while for the pulse it is not.

(2) Although w_1 for the pulse timing does not have a first moment of zero and therefore admits some drift, it admits considerably less than the step timing system.

(3) Figure 10.21 compares the frequency responses for the filters required for the step and pulse timing. The step in (b) requires the memory function in (d), this being the differential of the signal shape. The corresponding frequency response is then that shown in (f). But the pulse in (a) is the differential of the step in (b), so the frequency response in (e) is obtained by multiplying that in (f) by the frequency f. Thus the frequency response in (e) for the pulse timing falls to zero at $f = 0$, while that for the step does not. Thus, the step measurement is much more susceptible to low frequency $1/f$ noise.

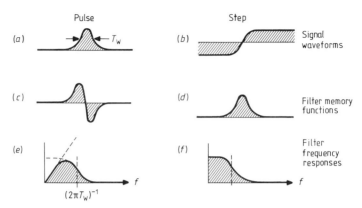

Figure 10.21 Diagram comparing the frequency responses of the filters required for step and pulse timing. (*a*), (*b*) Signal waveforms, (*c*), (*d*) filter memory functions, (*e*), (*f*) filter frequency responses.

(4) For the step timing system, a change in signal amplitude will cause a change in timing, since from figure 10.18(*c*) I_s will cross I_b at a different time. For the pulse timing, for which figure 10.6(*c*) is the corresponding diagram, $I_b = 0$ and so a change in I, caused by a change in signal amplitude, has no effect on the timing.

In the next section we show how the step timing method can be modified to overcome some of these problems. However, where it is possible to modify the experimental strategy so that the signal to be timed is a pulse rather than a step, this is probably the best solution.

10.5 OFFSET, DRIFT AND 1/f NOISE

We now consider in more detail the effects of the above factors in signal timing. It has just been established that, for a signal pulse, the optimum weighting for white noise tends to suppress these effects. Thus the main emphasis is upon the signal step.

Offset and drift
Figure 10.22 shows a method for removing offset and drift errors when timing a signal step. The central part of the weighting function w_1 in (*a*) and (*c*) is that which would be used for white noise. The 'wings' then constitute the usual modification for removing offset and drift, as developed in figure 9.16.

Figure 10.22(*d*) shows the output $\overline{v_{s1}}$ from the filter and compares it with the components $\overline{v_{sc}}$ and $\overline{v_{sw}}$ originating from the centre and wings separately. It is seen that the time at which $\overline{v_{s1}}$ crosses zero is also the time at which $\overline{v_{sc}}$ and $\overline{v_{sw}}$ cross each other. But at this time $\overline{v_{sw}}$ is equal to the mean of the initial and final

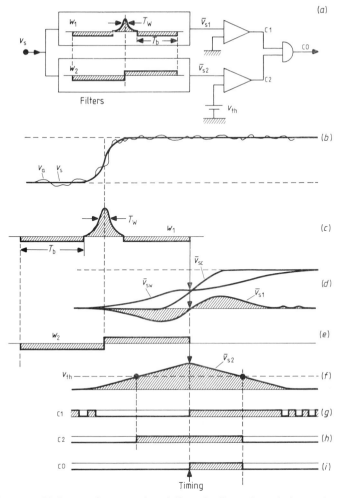

Figure 10.22 System for removing drift and offset when timing a signal step.

values of $\overline{v_{sc}}$ and this is the condition given in equation (10.22) as being that for optimum timing of the step.

The lower half of figure 10.22(a) is secondary to the main function and is the usual arrangement, corresponding to the lower half of figure 10.7(j), for avoiding false zero crossings in $\overline{v_{s1}}$ away from the signal step.

Compared with the more usual arrangement for step timing shown in figure 10.19, the present approach has one further advantage. The reference voltage v_b in figure 10.19 is replaced by $\overline{v_{sw}}$, which is proportional to the signal step amplitude. Thus, errors due to changes in the step amplitude are eliminated.

1/f noise

The modification to the weighting function shown in figure 10.22(c) when 1/f noise, rather than drift and offset, is present will now be determined. Figure 10.23(a) shows a suitable arrangement for timing a signal step in the presence of white noise. Here the noisy signal step v_a is synthesised using the impulse response of the filter F_s and the white noise generator. The required weighting function w' is then generated using the second filter F_s with a differentiator. The timing of the w' function is then adjusted until the output integral is zero and the timing is taken from w'.

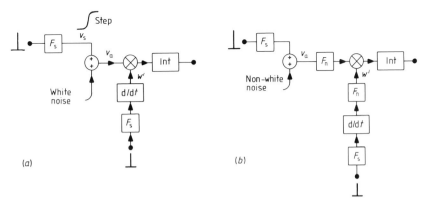

Figure 10.23 Optimum timing methods for a signal step in (a) white and (b) non-white noise.

Next, (b) shows the modifications needed when the noise is non-white. Now the pre-whitening filter F_n is needed for the signal and a similar filter is required in the w' arm to ensure that w' is still the differential of the signal input to the multiplier.

As usual, it is permissible to move all of the filters into one arm or the other, with suitable phase negation, to give either the transient generator or the transversal filter realisation. If the filters are moved to the signal arm to give the transversal filter realisation then the combination required is that in figure 10.24(a). Here, (b) shows the signal step and (c) the effect of the differentiator in (a) alone upon it. This will be the memory function corresponding to F_s^* and d/dt. For 1/f noise alone, the effect of the combined $F_n F_n^*$ filter is to multiply the frequency response by f while introducing no extra phase shift. The effect of this upon the function in (c) has already been established and produces the function in (d). This, then, is the combined memory function for the filter in (a) and represents the modification necessary to that in figure 10.22(c). Clearly the overall change is to reduce T_b to equal T_W.

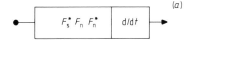

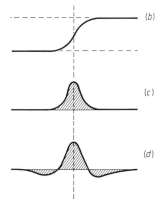

Figure 10.24 Transversal filter used in realising the system of figure 10.23(*b*). (*a*) Filter, (*b*) signal step, (*c*) F_s^* d/d*t* memory function, (*c*) $F_s^* F_n F_n^*$ d/d*t* memory function.

Pulse timing

Figure 10.25 shows the functions corresponding to those in figure 10.24 for the signal step, when a signal pulse is to be timed in the presence of $1/f$ noise. Since the signal pulse is the differential of the signal step it follows that each function in figure 10.25 will be the differential of that in figure 10.24. Thus, in particular, the final memory function of figure 10.25(*c*) for the pulse will be the differential of that in figure 10.24(*d*) for the step.

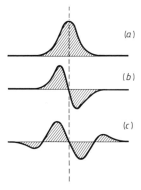

Figure 10.25 Transversal filter functions for a signal pulse in the presence of $1/f$ noise. (*a*) Signal pulse, (*b*) memory function for F_s^* d/d*t*, (*c*) memory function for $F_s^* F_n F_n^*$ d/d*t*.

As has already been pointed out, the frequency response corresponding to the unmodified function in figure 10.25(*b*) is that in figure 10.16(*b*) and this tends to reject $1/f$ noise. The only effect of adding the $F_n F_n^*$ modification is to multiply the frequency response by f. This makes the low frequency cut-off a second-order one, rather than the first-order cut-off of figure 10.16(*b*). This will have some small advantage in reducing the noise error, but only small. Notice also that the modified memory function of figure 10.25(c) has a first moment of zero, while that in (b) does not. This means that the modified system totally rejects constant rate drift while the unmodified system only suppresses it.

10.6 LOCK-IN AMPLIFIER

As a general rule, a single transient can only be timed with reasonable accuracy, or its amplitude measured, if the level of the noise when averaged over the period of the transient is less than the amplitude of the transient. Only then will the S/N ratio at the output of whatever process is applied to weight the signal optimally be large compared with unity. This is particularly the case with signal timing, where spurious zero crossings in the zero crossing method or spurious peaks in the peak detection method become increasingly difficult to distinguish from the true signal.

When the signal is periodic, however, the averaging time can be increased to cover many adjacent transients and much higher input noise levels can be accepted. Examples of this are the use of the PSD as described in chapter 7 for measuring the amplitude of a periodic signal buried in noise, and the phase-locked loop as described in §10.2 to time such a signal.

Figure 10.26 now shows an arrangement that combines the PSD and phase-locked loop methods, to allow both timing and amplitude measurements of a periodic signal that is submerged in noise. The VCO generates the waveform w which matches the known shape of the transient. The necessary function w' for the phase-locked loop, which is the differential of w, is then provided by the differentiator.

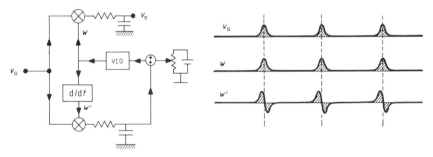

Figure 10.26 Lock-in amplifier. (a) System diagram, (b) waveform.

Here v_o represents the signal amplitude and the signal timing can be derived from the VCO output. Where the main interest is in the signal amplitude, the device is referred to as a 'coherent voltmeter' or a 'lock-in amplifier'.

10.7 AUTOCORRELATION METHODS

Sometimes it is required to measure the repetition period of a periodic waveform, the detailed shape of which is initially unknown. An example is the electrical ECG signal generated by the heart, which is shown in figure 10.27(a). Here the shape of the beat complex may be any of the variants shown in (a)–(c).

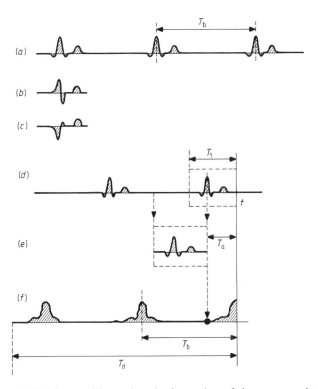

Figure 10.27 Diagram illustrating the formation of the autocorrelation function $\int_{t-T_1}^{t} v_a(t')v_a(t'-T_a)\,dt'$ as a means of determining the beat repetition period T_b of the ECG signal. (a) Waveform, (b) and (c) variants of beat complex in (a), (d) signal input, (e) weighting function derived from (d), (f) autocorrelation function.

Given the shape of the complex, a suitable method for locating each complex with minimum white noise error would be to use a weighting function equal to the complex shape and to use the above peak location method. That is, the weighting function is slid along the signal waveform and the time of occurrence of each complex is given by the time when the weighted sum of the signal waveform is maximum.

When the shape of the beat complex is initially unknown, the obvious solution is to determine it from the signal waveform. Figures 10.27(d)–(f) illustrate the procedure. The weighting function in (e) is derived from the signal waveform in (d). Then the weighted sum is formed for each value of the delay time T_a and the sum is plotted against T_a in (f). The beat period is then given by the position of the first peak in (f) as shown.

The function in (f) is termed the 'autocorrelation function' and the processor which generates the function an 'autocorrelator'. A suitable arrangement is shown in figure 10.28(a). Here, each running average filter holds the weighted sum of the input signal v_a for a different delay period T_a, up to the maximum of T_d, which is the total delay line period.

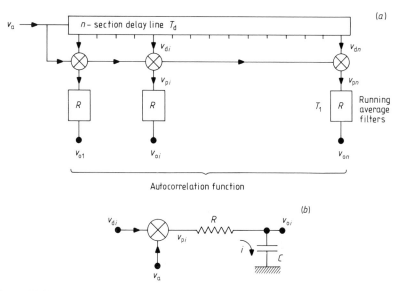

Figure 10.28 (a) Autocorrelator, (b) replacement of running average filter in (a) by CR filter.

The total amount of storage required for all of the running average filters is high, and this can be reduced by replacing each by the digital equivalent of the corresponding single-section CR filter. One of these, the ith, is shown in (b). Here v_{oi} represents the ith point on the autocorrelation function v_o. With the CR filter, only one variable needs to be stored for each filter and this is the capacitor voltage, which is also v_{oi}. Thus only two arrays are needed, the main shift register and v_o.

The procedure for updating v_o is then as follows. For the CR filter in (b) $i = C\,dv_{oi}/dt$ and $i = (v_{pi} - v_{oi})/R$ so $dv_{oi}/dt = (v_{pi} - v_{oi})/CR$.

Now, CR should equal the period T_1 for the running average filter. This gives the difference equation for the stepped digital process as $\Delta v_{oi}/\Delta t = (v_{pi} - v_{oi})/T_1$. This requires that the time increment Δt be sufficiently small to ensure that $\Delta t/T_1 \ll 1$.

Finally, since from (b) $v_{pi} = v_a v_{di}$, the procedure for updating v_{oi} is

$$v_{oi} \leftarrow v_{oi} + (v_a v_{di} - v_{oi}) \, \Delta t / T_1$$

and this must be executed for each of the v_{oi}.

Provided $CR = T_1$, the period of the beat complex, the effect of replacing the running average filters with the CR filters will be small. The effect is to weight further the weighted sum of v_a by the exponentially tapered memory function of the CR filter. Complications only arise when CR is made larger than the duration of the beat complex and approaches the beat repetition period T_b. Then the position of the peak on the autocorrelation function represents a merging of the values of T_b for adjacent beat repetition periods. This is equally true if T_1 is increased with the running average filters but with the CR filters the effect is always present to some small degree.

A disadvantage with the autocorrelation method is that it will only give the optimum timing error when the S/N ratio at the system input is high. Otherwise the weighting function derived from the signal is not the ideal one, being corrupted by noise.

The process is only that which gives the least noise error if the noise is white. When the noise is not white, this is simply rectified by using a pre-whitening filter. Note that this is not the filter $F_n F_n^*$ used with earlier methods. This overcorrects the non-white noise, and to make the noise white requires only one part of the filter, say F_n. Again unlike the previous methods, the phase characteristic of this pre-whitening filter is of little importance. This is partly because the detailed shape of the signal is not known, and partly because the shape of the weighting function is always the correct one for the signal.

It is also advantageous, when the upper limit of the signal spectrum is known, to remove all higher frequency noise components at the system input. This improves the chances of ensuring that the input S/N ratio is high.

A minor practical complication in using the autocorrelator can be seen from figure 10.27(d). Imagine that the 'present' time t shown is further ahead by T_1. Now, the weighting function derived from the signal is zero and the autocorrelation function in (f) will also be zero. Thus, when T_1 is comparable with the beat complex duration, as shown, the autocorrelation function peaks move up and down in amplitude with each beat. Clearly the best time to measure T_b for each beat is when the peaks are maximum. This problem only arises when true beat-to-beat resolution of T_b is required. If only an average is required over several beats, T_1 or CR can be made longer. This makes the weighting function w span several beats, and so w is never zero, and the pulsation in the autocorrelation function peaks is much reduced.

Appendix: Fourier analysis

In chapter 4 it was stated that any periodic waveform of period T_0 can be resolved into a series of sine wave components of frequency $0, f_0, 2f_0, 3f_0$ and so on, where $f_0 = T_0^{-1}$. The proof that this is so is beyond the scope of the present text and the object of this appendix is to show, assuming the above statement to be true, how, given the periodic waveform, the magnitudes and phases of the sine wave (Fourier) components can be calculated.

Let v be the periodic waveform. Then it is assumed that v can be expressed in the form

$$v = A_0 + \sum_{n=1}^{\infty} [A_n \cos(n\omega_0 t) + B_n \sin(n\omega_0 t)] \qquad (A.1)$$

where $\omega_0 = 2\pi f_0$ and n is an integer.

The problem then is to calculate the A and B coefficients. Consider the effect of forming the integral $\int_0^{T_0} v \cos(m\omega_0 t)\, dt$ where m is an integer. From equation (A.1)

$$\int_0^{T_0} v \cos(m\omega_0 t)\, dt = \int_0^{T_0} A_0 \cos(m\omega_0 t)\, dt + \sum_{n=1}^{\infty} \left(A_n \int_0^{T_0} \cos(n\omega_0 t)\, dt \right.$$

$$\left. + B_n \int_0^{T_0} \sin(n\omega_0 t)\cos(m\omega_0 t)\, dt \right). \qquad (A.2)$$

Each integral on the right-hand side is zero except the term $A_n \int_0^{T_0} \cos(n\omega_0 t)\cos(m\omega_0 t)\,dt$ for which $m = n$. This term has the value $A_n T_0/A$. Thus

$$A_n = 2T_0^{-1} \int_0^{T_0} v \cos(n\omega_0 t)\, dt. \qquad (A.3)$$

By a similar argument

$$B_n = 2T_0^{-1} \int_0^{T_0} \sin(n\omega_0 t)\, \mathrm{d}t. \tag{A.4}$$

Finally, if $\int_0^{T_0} v\, \mathrm{d}t$ is formed for both sides of equation (A.1), then every term of the right-hand side is zero except the first, giving

$$A_0 = T_0^{-1} \int_0^{T_0} v\, \mathrm{d}t. \tag{A.5}$$

Thus, using equations (A.3) to (A.5), all of the A and B coefficients can be calculated.

As an example let us calculate the coefficients for the square wave of figure 4.1, thereby justifying equation (4.1). From the waveform $\int_0^{T_0} v\, \mathrm{d}t = 0$ giving $A_0 = 0$. From equation (A.3)

$$A_n = 2T_0^{-1} \left(\int_0^{T_0/2} A \cos(n\omega_0 t)\, \mathrm{d}t + \int_{T_0/2}^{T_0} (-A)\cos(n\omega_0 t)\mathrm{d}t \right). \tag{A.6}$$

Here each term of the right-hand side is equal to 0, so $A_n = 0$ for all n.
From equation (A.4)

$$B_n = 2T_0^{-1} \left(\int_0^{T_0/2} A \sin(n\omega_0 t)\, \mathrm{d}t + \int_{T_0/2}^{T_0} (-A)\sin(n\omega_0 t)\, \mathrm{d}t \right). \tag{A.7}$$

For even values of n this is zero and for odd values

$$B_n = 4A/(n\pi). \tag{A.8}$$

Thus v may be expressed in the form

$$v = \frac{4A}{\pi} \sum_{n=1}^{\infty} n^{-1} \sin(2\pi f_0 n t)\, \mathrm{d}t \tag{A.9}$$

where n is an odd integer.

This is identical to the Fourier series of equation (4.1), except that there r is used as the index, and thus the equation is justified.

Index